Palgrave Studies in Agricultural Economics and Food Policy

Series editor
Christopher Barrett
Cornell University
Ithaca, New York, USA

Agricultural and food policy lies at the heart of many pressing societal issues today and economic analysis occupies a privileged place in contemporary policy debates. The global food price crises of 2008 and 2010 underscored the mounting challenge of meeting rapidly increasing food demand in the face of increasingly scarce land and water resources. The twin scourges of poverty and hunger quickly resurfaced as high-level policy concerns, partly because of food price riots and mounting insurgencies fomented by contestation over rural resources. Meanwhile, agriculture's heavy footprint on natural resources motivates heated environmental debates about climate change, water and land use, biodiversity conservation and chemical pollution. Agricultural technological change, especially associated with the introduction of genetically modified organisms, also introduces unprecedented questions surrounding intellectual property rights and consumer preferences regarding credence (i.e., unobservable by consumers) characteristics. Similar new agricultural commodity consumer behavior issues have emerged around issues such as local foods, organic agriculture and fair trade, even motivating broader social movements. Public health issues related to obesity, food safety, and zoonotic diseases such as avian or swine flu also have roots deep in agricultural and food policy. And agriculture has become inextricably linked to energy policy through biofuels production. Meanwhile, the agricultural and food economy is changing rapidly throughout the world, marked by continued consolidation at both farm production and retail distribution levels, elongating value chains, expanding international trade, and growing reliance on immigrant labor and information and communications technologies. In summary, a vast range of topics of widespread popular and scholarly interest revolve around agricultural and food policy and economics. The extensive list of prospective authors, titles and topics offers a partial, illustrative listing. Thus a series of topical volumes, featuring cutting-edge economic analysis by leading scholars has considerable prospect for both attracting attention and garnering sales. This series will feature leading global experts writing accessible summaries of the best current economics and related research on topics of widespread interest to both scholarly and lay audiences.

More information about this series at
http://www.springer.com/series/14651

May T. Yeung · William A. Kerr
Blair Coomber · Matthew Lantz
Alyse McConnell

Declining International Cooperation on Pesticide Regulation

Frittering Away Food Security

palgrave
macmillan

May T. Yeung
Johnson Shoyama Graduate School
of Public Policy
University of Saskatchewan
Saskatoon, SK
Canada

William A. Kerr
Department of Agricultural and
Resource Economics
University of Saskatchewan
Saskatoon, SK
Canada

Blair Coomber
Ottawa, ON
Canada

Matthew Lantz
Bryant Christie Inc.
Seattle, WA
USA

Alyse McConnell
Bryant Christie Inc.
Sacramento, CA
USA

Palgrave Studies in Agricultural Economics and Food Policy
ISBN 978-3-319-60551-7 ISBN 978-3-319-60552-4 (eBook)
DOI 10.1007/978-3-319-60552-4

Library of Congress Control Number: 2017944113

Cover illustration: Pattern adapted from an Indian cotton print produced in the 19th century

Printed on acid-free paper

This Palgrave Macmillan imprint is published by Springer Nature
The registered company is Springer International Publishing AG
The registered company address is: Gewerbestrasse 11, 6330 Cham, Switzerland

Jeff and Suzy Lantz
Nicholas Perdikis
In loving memory of Manabu Donald Nakahama, Jake and Max

Series Editors' Foreword

The global food system faces many changes and challenges in the coming years and decades. This important new volume on international cooperation around pesticide regulation in food trade speaks directly to the intersection of three of those important challenges.

The first revolves around growing consumer concerns for the natural environment and for human health. In most countries, a government agency must approve the sale and use of agro-chemicals and other regulated chemical products, as well as the maximum allowable limits of pesticide residues permitted on imported foodstuffs. The purpose of government regulation is to safeguard food quality and to protect the environment against potential harm while simultaneously ensuring fair competition with domestic producers. Controls over pesticides, most commonly in the form of restricting use and setting maximum residue limits, have grown steadily since World War II. Over the past two decades, private firms have begun to set their own standards that are stricter still, catering to growing consumer demand for environmental and health attributes of foods. However, as political and market demands vary among countries, controls over pesticides have evolved differentially around the globe. These variances become both larger and more troublesome for cross-border trade.

The second big change concerns the growing importance of international trade in food, especially with anticipated future growth in food demand heavily concentrated in African and Asian countries that are currently net food importers. The prospect of greater country-specific supply fluctuations due to climate change further increases the importance of orderly agricultural trade, as Kym Anderson's (2016) excellent volume in this series explains. As international trade in food becomes ever more important for advancing food security objectives and helping the world adapt to the prospect of more frequent extreme weather events, barriers to trade require increased attention.

Third, climate change and increasing trans-boundary transport will also shift the geographic ranges of various pests and pathogens and facilitate greater incursions by invasive species into agro-ecosystems. These changes will necessitate protective response by agricultural producers, some of which will inevitably come in the form of changing, perhaps even expanding, use of pesticides, and other agro-chemicals. Especially, if resistance to genetically modified organisms continues in many countries, pesticides' role in protecting global agricultural production will almost surely rise, as Matin Qaim's (2016) outstanding study in this series explains.

The confluence of these three phenomena—growing consumer pressure for environmental and health standards in foods, expanding international trade, and shifting geographies of pests that will change and potentially expand pesticide use—makes international cooperation around pesticide regulation ever more important. Yet, as with many other regulatory barriers to trade, cooperation can prove difficult and elusive, especially with inadequate careful study to inform deliberations and agreements.

Especially as compared to other trade barriers, such as tariffs, quotas, and export subsidies, or other modern agricultural technologies such as genetically modified organisms, relatively a little attention has been paid to regulatory trade barriers associated with maximum residue limits for pesticides on imported food products. As May Yeung, William Kerr, Blair Coomber, Matthew Lantz, and Alyse McConnell explain in this carefully researched and reasoned new volume, this issue merits far more careful study, extensive discussion, and prompt action. The problem of asynchronous maximum residue limits—i.e., limits that differ among trading parties—is growing and increasingly interfering with orderly agricultural trade. This will become more problematic as other shifts within the global food system make efficient and rapid trans-boundary food shipments increasingly valuable.

The book's authors skillfully unpack the complex relationship among the science, politics, and economics of pesticide regulation, and how these affect international trade. They devote considerable attention to the motivations of different actors within this multinational system and the resulting political economy of initiatives to harmonize maximum residue limits internationally. I found the work extremely enlightening.

It is a great pleasure to include this outstanding new book in the Palgrave Studies in Agricultural Economics and Food Policy series. It will prove an essential reference about an important but oft-overlooked topic that is bound to grow in importance for food policy in the early twenty-first century.

Christopher Barrett
Cornell University

References

Anderson, Kym. 2016. *Agricultural Trade, Policy Reforms, and Global Food Security*. New York: Palgrave Macmillan.

Qaim, Matin. 2016. *Genetically Modified Crops and Agricultural Development*. New York: Palgrave Macmillan.

Preface

The main concerns regarding barriers to trade have historically been that they reduce opportunities for firms wishing to engage in international commercial transactions, inhibit the efficient use of resources, and reduce the welfare of societies able to engage in trade. The agri-food sector remains one of the sectors where barriers to trade abound. Trade barriers can arise from two major sources. The first, and the one that has received the most attention, is those barriers put in place by governments responding to requests for economic protection from domestic firms. The second arises in the form of, often unintentional, friction in the trading system. This friction raises the costs for firms, thus reducing incentives to engage in international trade. While unintended, these barriers to trade can often be mitigated by proactive activities of governments either singly or acting together through international fora. A failure to act in these circumstances can have just as significant an effect on trade and societal welfare as the purposeful erection of trade barriers motivated by a desire to provide economic protection.

While the problem with unintended barriers to trade has a long history, what is new in the case of trade in agri-food products is the looming threat to global food security. Global population is forecast to increase from 6.7 billion people to 9 billion by 2050. This is a 35% increase in less than 35 years. Furthermore, rising incomes among a considerable proportion of the globes' poorest consumers will mean increased demand for food. The combination of more mouths to feed and rising incomes will mean a doubling of food demand over this short time-span. At the same time, the rate of increase in agricultural productivity is slowing and climate change is likely to adversely affect agricultural output in areas of the world where food is most needed. The bottom line is that increasing agricultural production sufficiently to prevent a deterioration in food security represents a major challenge for the first half of the twenty-first century.

While efforts to increase local production will form a major component of achieving food security goals, in many areas of the world, this strategy will not be successful in terms of being able to maintain levels of food security. Much of the growth in population is slated to take place in areas of the world where agricultural production levels are already not environmentally sustainable. Attempts to increase production will lead to erosion of the agricultural resource base and declining food security. Hence, the additional supplies required to maintain food security will have to come through international shipments from food surplus areas to food deficit areas. Any friction in the trading system that inhibits the movement of agri-food products internationally in a systematic way warrants investigation as to its causes and how its effects can be mitigated.

One of the frictions in the international movements of agri-food products coming to the forefront of concern is the increase in asynchronous maximum residue limits (MRL) for pesticides on imported food products. Divergence in these standards leads to the turning back of cross border food shipments in the short run and inhibits investment in food trade infrastructure over the longer run. While other instances of asynchronous regulatory regimes, such as those governing genetically modified organisms, have received considerable attention, the issue of MRLs has not. Given that these chemicals are used to enhance agricultural production on a truly global scale, the impact of trade restrictions due to regulatory divergence in this area can be extensive and has the potential to make the achievement of food security goals going forward much more difficult. Furthermore, the problem of asynchronous MRLs is worsening.

This serious and deteriorating situation provided the motivation for undertaking this research. It is a clear case of governments either being relatively indifferent to the problem or unable to effectively coordinate their efforts to solve the problem—because it could be solved. The book delves into the source of the problem, the role of MRLs in domestic agricultural production and trade in food products, the motivations of governments and the incentives for policy makers and administrators, and international initiatives to harmonize MRLs internationally, and provides suggestions as to how to move forward to resolve the issue.

Funding from the Canola Council of Canada to support this research is gratefully acknowledged.

Saskatoon, Canada	May T. Yeung
Saskatoon, Canada	William A. Kerr
Ottawa, Canada	Blair Coomber
Seattle, USA	Matthew Lantz
Sacramento, USA	Alyse McConnell

Contents

Abbreviations

APEC	Asia Pacific Economic Cooperation
ASEAN	Association of South East Asian Nations
BSE	Bovine Spongiform Encephalopathy
CAC	Codex Alimentarius Commission
CCPR	Codex Committee on Pesticide Residues
Codex	Codex Alimentarius Commission
EFSA	European Food Safety Agency
EPA	Environmental Protection Agency (US)
EU	European Union
FAO	Food and Agriculture Organization
FDA	Food and Drug Administration (US)
FSCF	Food Safety Cooperation Forum (APEC)
GAP	Good Agricultural Practice
GJR	Global Joint Review
GLP	Good Laboratory Practice
HIT	Heterogeneity Index Trade Regulations
ICGCC	International Crop Grouping Consulting Committee
IPPC	International Plant Protection Convention
IR-4	Inter-Regional Research Project #4
JMPR	Joint Meetings on Pesticide Residues
LoD	Level of Detection
MHLW	Ministry of Health, Labor and Welfare (Japan)
MRL	Maximum Residue Limit
NTB	Non-Tariff Barrier
OECD	Organization for Economic Cooperation and Development

OIE	World Organization for Animal Health (formerly the Office International des Epizooties)
PFSS	Pesticide Food Safety Standard
PMRA	Pest Management Regulatory Agency (Canada)
ppm	parts per million
RASFF	Rapid Alert System for Food and Feed (EU)
SPS	Sanitary and Phyto-Sanitary
TBT	Technical Barriers to Trade
TFDA	Taiwan Food and Drug Agency
UK	United Kingdom
UN	United Nations
US	United States
vCJD	variant CreutzfeldtJakob
WHO	World Health Organization
WRF	Wine Regulatory Forum
WTO	World Trade Organization

List of Figures

List of Tables

CHAPTER 1

Why Maximum Residue Limits for Pesticides are an Important International Issue

Abstract The food security challenges that will arise over the next 35 years are outlined. The importance of international trade in meeting food security goals is explained. The threat to trade in agri-food products posed by increasingly asynchronous import regulations for maximum residue limits (MRL) for pesticides is presented. The problems facing MRL harmonization are detailed.

Keywords Global food supply · Pesticide residues · International agricultural trade · Differing MRLs · Missing MRLs · Misaligned regulatory systems

1.1 Food Security and Maximum Residue Limits for Pesticides

By 2050—a little over 30 years from now—the world's population is projected to grow from approximately 6.7 billion people to 9 billion people. This represents a 35% increase in the number of mouths needing to be fed. It represents a major challenge for the global food system to ensure that this extra demand for food can be met without a decline in food security for some segments of the global community—likely the world's poorest inhabitants.

In addition to the *extra mouths to feed* by 2050, a significant proportion of those who have been among the world's poorest are seeing their

M.T. Yeung et al., *Declining International Cooperation on Pesticide Regulation*, Palgrave Studies in Agricultural Economics and Food Policy, DOI 10.1007/978-3-319-60552-4_1

incomes rise. As incomes of the poor rise, most of that extra income is spent on food. This gives a further boost to food demand, such that the combination of additional population and rising incomes is projected to double the global demand for food by 2050 (Evans 2009).

Doubling global food supply by 2050 is a herculean task. It is made more difficult, because most of the growth in population and rising incomes of the very poor will take place in developing countries, where agricultural systems are already stressed and not being farmed in a sustainable fashion. They are also often in parts of the world where the effects of climate change on agricultural production are expected to be the greatest (Kerr 2011). Furthermore, almost all of the population growth is projected to arise in cities, meaning that the extra population will not be engaged in food production. While increased production from agriculture in these areas will certainly contribute to meeting the rising demand for food, it is unlikely to be able to satisfy all the additional need for food. Hence, the shortfall will have to be filled through international trade (Kerr 2011).

Most of the areas of the world capable of producing additional food surplus to domestic needs are in developed countries, in the Black Sea region or South America (Burkitbayeva and Kerr 2015). While significantly increasing agricultural exports in these areas has its own set of agronomic, political, and logistical challenges (Kerr 2014), they do present considerable potential contributions to realizing future global food security goals. One of the major potential constraints on this contribution that food surplus areas can make towards putting additional food onto the international market is trade barriers. Trade barriers have two effects. In the first instance, they directly reduce the quantities of food exported. Second, they reduce profitable trade opportunities, thus discouraging investments in projects designed to increase food production. In short, trade barriers act as a brake on increasing agricultural productivity. Increasing the productivity of agriculture is essential to the task of improving food security. Thus, as Smyth et al. (2015, p. 17) state:

> Rather, future food security requires that we look at all the impediments to higher agricultural productivity …

One of the current and rising impediments to agricultural trade is differing levels for pesticide residues on food production among countries. Given the preponderant use of pesticides in agricultural production globally, trade restrictions arising from lack of harmonization in pesticide regulations have the potential to lead to wide ranging trade inhibiting effects. This problem provides the motivation for this book. It examines the issues around pesticide residues and their effects on trade. It makes the case that this major inhibitor of international agricultural trade represents a needless *frittering away of food security.*

Maximum Residue Limits[1] (MRL) for pesticides is one of the most difficult challenges facing the international trade of agricultural products. It is a global issue affecting farmers and exporters who find that if they correctly follow the instructions on a crop protection product's label, meaning that they have a perfectly legal crop in their home country, it can still be rejected upon arrival in a foreign market. Rejection can be due to a residue level that exceeds a foreign market's established MRL or because the MRL has not been established for that market. The rejection scenario is sufficiently pervasive and repetitive that it has become a major trade barrier, but unlike other agricultural trade barriers, MRL issues are largely unintentional. In other words, they are not the result of economic protectionism. They result instead from internationally misaligned regulatory systems. Unlike the case with tariffs, quarantine issues, or other non-tariff barriers, one rarely, if ever, hears of a domestic agriculture industry lobbying its government to not establish a MRL, because it wants to keep a foreign commodity out of the market. Instead, MRLs are simply either missing or established at levels different from the standards of other countries.

Even if unintentional, misaligned MRLs have significant trade implications. In 2015, for example, hundreds of millions of dollars of United States (US) agricultural exports were put at risk due to differing interpretations of MRLs between, for example, the US and the European Union (EU). In other cases, Philippine bananas experienced a major decline in sales in South Korea due to a MRL regulatory change. On a smaller scale, numerous individual shippers receive calls that their container of fruit has been rejected in a foreign port, because it tested positive for a residue of a chemical-based product that had not yet been approved in that importer's market. The rejected shipment can either be rerouted or destroyed. Government officials, from embassy staff to senior-level diplomats, are being increasingly asked to deal with MRL problems. To the frustration of farmers, who want to use the latest crop protection product, and to pesticide registrants, who have spent millions developing that product, growers are being told by buyers, *do not use that compound, there is no MRL.*

How did the system for international movement of agricultural products arrive at this relatively dysfunctional outcome? 20 years ago, MRLs were not a major issue in international agricultural trade. Some countries had MRL lists and many countries had a system for registering pesticides for domestic usage, but pesticide residue testing on imported product was not considered a priority. Countries would defer to international Codex[2] MRLs or an exporting country's MRL regulations if there was a discrepancy, and there were not many.

This trade-friendly system began to change in the late 1990s. In 1999, Taiwan announced that it would establish its own *positive*[3] MRL list and imports would need to meet those standards to enter the market. This caused concern among exporters shipping to Taiwan as the Taiwan MRL list was significantly shorter than other country MRL lists, including that of the US. Senior-level diplomatic negotiations occurred to ensure that trade was not halted as a result of Taiwan's new system.

Additional countries began to overhaul their MRL systems as they pertained to imports. After a series of food scares in the early 2000s, Japan announced its intention to revise and significantly expand its own list of national MRLs. That list went into effect in May 2006. The EU decided to consolidate its differing member state MRLs into a unified community-wide list in 2008. During the same period, after a series of regulatory reforms allowed for a more rapid process to establish MRLs, Canada elected to significantly increase the number of MRLs in its system. After 2010, Australia also expanded its MRL list after creating a new system for the setting of import tolerances, especially if Codex MRLs existed. On August 1, 2014, Hong Kong's positive list MRL system went into effect. South Korea is currently proposing to eliminate its current system that uses a combination of national MRLs, Codex MRLs, and deferral paths to allow only Korean national MRLs by January 1, 2019 and earlier for certain commodities. China and India are examining significant expansions of the number of MRLs in use in their markets.

Why have these individual countries decided they need their own MRL systems? Several factors have contributed. All have expressed a desire to update older food safety regulations. Many have faced growing civil society interest in food safety after food scares. They have determined that a modern food safety system that relies on their own national standards, as opposed to either not having defined standards or deferring to an external standard, is needed. Differing diets in Asia are sometimes cited as a justification for a national standard. Some countries, especially in Asia, where many of the new national positive lists are being developed, are no doubt observing neighbors establishing national MRL systems and in a cascade effect, saying, *we need our own*. How these systems are put in place and implemented can cause significant challenges for the importation of food.

The consequence of these developments is that farmers and exporting firms must be aware of a variety of MRLs when growing and shipping their agri-food product. Moreover, pesticide registrants must now have a global plan to obtain MRLs in numerous markets if they expect to be able to sell their crop protection products. This may also result in increasing economies of scale for MRLs for larger acreage crops, however, smaller acreage, or specialty crops face delays or completely lack MRLs as they do not present the economies of scale that make seeking a MRL economically viable. This is recognized problem that is being addressed in some countries with initiatives such as the IR-4 Project[4] in the US.

1.2 The Trade Challenges Posed by MRL Asynchronicity

The current state of regulatory divergence among countries pertaining to MRLs presents a variety of challenges for those wishing to engage in agricultural trade, including:

1.2.1 Differing MRLs

With multiple national MRL systems, the result is that MRLs often differ between markets. Different markets have different tolerances for risk and differing evaluation systems, which can result in an array of varying MRLs. The EU has, for example, often established MRLs at lower (i.e., stricter) levels than the US. Growers must be aware of differing MRLs and determine if a more restrictive MRL in an export market makes their shipments problematic when it comes to accessing that market. Residues should come in well below an established MRL, but if a foreign MRL is, for example, half of a domestic MRL, there may well be a problem related to market access.

1.2.2 Missing MRLs

Even with national MRL lists, the number of MRLs established in a country can vary significantly. The US, for example, has over 20,000 unique MRLs, while most other countries have a much smaller number. A missing MRL is a particular challenge for those attempting to export as the consequences are not clear. If there is a missing MRL, does the country have a default tolerance, and if so does that default level apply? If not, does the potential importing market test for the compound? If so, there is a possibility of a violation. What are the imposed sanctions for a violation? Is it possible to obtain a MRL for the compound in question? Does the country just outright reject the shipment? Such lack of transparency increases the risk associated with exporting, sometimes to the point of deterring attempts to export entirely.

1.2.3 Sanctions Policies

The action which a country takes when a residue violation occurs can significantly alter the disruptiveness arising from differing MRL levels. Japan reserves the right to increase testing on all shipments of a commodity after two residue violations. South Korea can seek increased testing on a commodity for 5 years for a single infraction. Taiwan and the EU publish violations on websites. At times, violations are covered in the media. Other markets take action against individual shippers only, and some do not test at all. Differing sanctions can determine how challenging MRLs are for a foreign exporter in a particular market.

1.2.4 Misperceptions of Food Safety Issues

One of the most challenging aspects of MRLs is the erroneous assumption made by some members of civil society that MRLs represent actual food safety standards, meaning that a residue found above a MRL places the consumer at significant risk of becoming ill. Although MRLs are designed to keep consumers safe, they are much stricter than any level of potential toxicity for human health or the environment. MRLs are intended to be used as signals to demonstrate that Good Agricultural Practices (GAP)[5] are being followed in agricultural production. If a chemical was applied according to the label, the residue should be below the MRL. If it exceeds the MRL, something untoward happened.

The breach of the limit can be investigated, but it does not mean that there is a potential risk to human health.

1.2.5 *The Politics of Import Tolerances*

Given the need to have MRLs established in foreign markets to ensure that exports are compliant with importing country standards, there is often a desire to have a commercially viable import tolerance established. Import tolerances allow residue levels on imported food for crop protection products that are not registered for use in the destination market. The import tolerance allows an exporter to use a crop protection product and be compliant in the foreign market. These can be controversial, because the MRL is established, but it does not mean that farmers in the importing market can use the product domestically on their crops. This can sometimes prove a political challenge. In Canada, the EU, Taiwan, and South Korea, producers have complained that their regulators should not be establishing import tolerances for imported food, and instead, should focus limited resources on fully registering new chemicals and uses within their home market for their own farmers. Domestic producers only want to see MRLs established with full registrations, not import tolerances. Meanwhile, some major import markets, such as China, do not currently have a system in place to establish import tolerances for residues of crop protection products not registered for use in China. China is considering to adjust this policy in the future, but currently, the only way to obtain a MRL in China is through a full domestic registration of the crop protection product, complete with trials within China. Any change will require several years.

1.2.6 *Older Products/Generics*

When a market announces an overhaul of its regulatory system requiring the establishment of new MRLs, doing so for older generic[6] compounds can prove challenging. If a country seeks submissions of data packages, as Korea is currently doing, obtaining that data and determining who would submit and pay for such a review on older compounds pose a significant obstacle. Post-patent materials are typically sold by multiple registrants who may not have the original data associated with the compound. They also may not have offices in the foreign market or the capacity to make a submission. At the same time, the companies that generated the data years ago have little financial incentive to work on a foreign MRL approval for a product which they no longer receive patent protection for and/or sell. Hence, use of these generic products, which are commonly used by farmers due to their affordability and effectiveness, can be inhibited by the lack of a MRL. A case study discussing the issue of generics and older products is provided in Chap. 6.

1.2.7 Private Sector Standards

Private sector standards pertaining to MRLs have also emerged, which pose a further challenge for those interested in exporting. These private standards have become particularly prevalent across Europe. Governmental MRLs are based on official regulatory reviews. Although these reviews may differ from country to country, they are science-based. For private sector standards, some retail food chains have announced that any products in their stores will be a certain percentage below the country's existing official MRLs. They also can require that certain chemicals, although legal to use and having established MRLs, not be applied to produce or agri-food products to be sold in their stores. These private standards may not be based on science, but instead are driven by attempts to increase profits through product differentiation (i.e., strategic marketing) and, hence, pose further challenges for potential exporters.

MRL regulatory harmonization is a complex, technical problem, with multiple contributory factors affecting most aspects of agri-food production and trade. This book will assess potential opportunities that can contribute towards mitigating the trade impeding effects of misaligned global MRLs.

Notes

1. Also known as maximum residue levels or tolerances.
2. Codex refers to the Codex Alimentarius Commission which is the United Nation's food safety organization. It is discussed in detail further later in the book.
3. A positive list is one where items included on the list are affected by the related change in contrast to a negative list where items listed are excluded from related changes.
4. The IR-4 Project is a resource for supplying pest management tools for specialty crop growers by developing research data to support new EPA tolerances (MRLs) and labeled product uses. It is discussed in greater detail later in the book.

5. The FAO defines Good Agricultural Practices as "practices that address environmental, economic and social sustainability for on-farm processes and result in safe, quality food and non-food agricultural products" (FAO 2008).
6. A generic compound is one that is no longer under patent and can therefore be produced and marketed by firms other than the original patent holder.

References

Burkitbayeva, S., and W.A. Kerr. 2015. Accession of KRU to the WTO: The Effect of Tariff Reductions on KRU and International Wheat Markets. In *Transition to Agricultural Market Economies, The Future of Kazakhstan, Russia and Ukraine*, ed. A. Schmitz, and W. Meyers, 183–190. Wallingford: CABI.

Evans, A. 2009. *The Feeding of the Nine Billion: Global Food Security*. London: Chatham House.

FAO. 2008. *Good Agricultural Practices*. Food and Agriculture Organization of the United Nations. http://www.fao.org/prods/gap/.

Kerr, W.A. 2011. The Role of International Trade in Achieving Food Security. *Journal of International Law and Trade Policy* 12 (2): 44–53. http://ageconsearch.umn.edu/bitstream/117818/2/kerr12-2.pdf.

Kerr, W.A. 2014. Food Security and Trade: Some Supply Conundrums for 2050. *Journal of International Law and Trade Policy* 15 (2): 115–132. http://law.usask.ca/documents/estey-journal/Kerr15-2lay.pdf.

Smyth, S.J., P.W.B. Phillips, and W.A. Kerr. 2015. Food Security and the Evaluation of Risk. *Global Food Security* 4: 16–23.

CHAPTER 2

The Importance of Trade for Food Security

Abstract The important contribution international trade has in meeting the food security challenges that will arise in the first half of the twenty-first century is explained. The dual effect of rising population and increasing incomes among the world's poorest consumers on food demand is outlined. Slowing rates of increase in global agricultural productivity will add to the difficulties in increasing food production. The areas where food demand will increase the most are not where increases in food production can take place, meaning that international movement of foodstuffs must fill the gap.

Keywords Food demand · Food security · Population growth · Food production · Agricultural research · Productivity

Food security is set to become a major global challenge as the world's population increases to 9 billion by 2050. This one-third increase in the number of people that need to be fed, combined with rising incomes among a considerable proportion of the world's poorest inhabitants, will stretch the capacity to produce food to its breaking point. A failure in the ability to increase food production at a sufficient pace to keep up with rising demand can only mean a decline in the food security status of the world's poorest citizens. The fragility of the world's food system came into sharp focus during the recent food crisis when there was a major

M.T. Yeung et al., *Declining International Cooperation on Pesticide Regulation*, Palgrave Studies in Agricultural Economics and Food Policy, DOI 10.1007/978-3-319-60552-4_2

spike in the prices of staples in 2007–2008. The crisis stemmed from a combination of multiple harvest reducing events around the world and trade-restricting practices of some major exporting countries (Giordani et al. 2012). The resulting spike in food prices pushed over 75 million additional people into extreme hunger and malnutrition (FAO 2008; World Bank 2009). This was with current population levels and points to the need for food's international trade system to work with a minimum of trade restrictions (Kerr 2011a). Food security has three fundamental components: (1) food production[1] must increase at least at the pace of increasing demand for food; (2) food must be consistently available; and (3) individuals must have sufficient resources to acquire food. One is not food secure if markets or supermarkets are fully stocked with food, but one does not have sufficient funds to purchase it. This is the case, for example, for poor individuals in developed countries where food stores are well stocked, but incomes are not sufficient to purchase it in quantities that will provide adequate nutrition over time. Such individuals are not food secure. It is why there is a need for *food banks* and other food sources for the poor such as *soup kitchens.*

Famines can also arise from lack of resources among the poor (Kerr 2011a). A famine usually starts with some shock to agricultural production which reduces available food. This causes prices to rise. If the trade infrastructure or political situation does not allow for an inflow of food from other areas in response to the rise in food prices, then the poor quickly exhaust their resources attempting to acquire what food is still on offer. Once they have exhausted their resources, starvation follows—even if adequate food supplies subsequently become available.

This was the case, for example, in the Irish potato famine—the *anGortaMór*—of the late 1840s. According to Bloy (2002, p. 1):

> The Irish crisis was used as an excuse by Peel in order for him to the repeal the Corn Laws in 1846, but their removal brought Ireland little benefit. The major problem was not that there was no food in Ireland — there was plenty of wheat, meat, and dairy produce, much of which was being exported to England — but that the Irish peasants had no money with which to buy the food. The repeal of the Corn Laws had no effect on Ireland because however cheap grain was, without money the Irish peasants could not buy it.

It was not until the British government made resources available to establish soup kitchens that some of the poor were able to escape starvation. The efforts of the *World Food Program* provide a contemporary example (Kerr 2015).

If most individuals have adequate resources, such that under normal circumstances, they can acquire sufficient food, they are still not food secure if food is not available—almost on a daily basis.[2] One only has to observe the reaction of individuals in developed countries when a hurricane, flood, or other forecast natural disaster can be expected to disrupt deliveries of food—even for a few days. The result is a rush on food retailers with shelves being emptied and, sometimes being looted, in panic. It is clear that those engaging in these types of activities do not feel themselves to be food secure.

Much of the world's lack of food security arises when, for a variety of reasons, potential food supplies are not able to respond, through internal or international trade, to the incentives provided by price increases in the wake of a disruption to agricultural production. The reasons for a failure to capitalize on potential arbitrage opportunities include lack of transportation infrastructure (roads, ports, and trucks), lack of communication networks, such that information on relative prices is not available or government restrictions on movements of food—including asynchronous regulations—as well as wars and insurrections.

In modern market economies, there are local disruptions to food production all the time—drought, flooding, frost, diseases, pests, etc., still decimate local production on a regular but unpredictable basis. Food retailers (and/or other players in food supply chains) are, however, easily able to identify alternate sources of supply for the product, or its close substitutes. Communication and transportation infrastructure function well, and there is a minimum interference in the movement of food. The result, from the perspective of consumers, is a seamless shift in supply with minimal effect on prices, such that they may not even know about the disruption to production. Seldom does one go to the supermarket and what one wants[3] is not there. This ability to arbitrage is a hallmark of food security in modern market economies.

In some developing countries, markets may not work as seamlessly to mitigate the effects of local food supply disruptions. As a result, food security suffers. While the rich are always likely to have the ability to acquire food,[4] the poor's food security will decline. In some cases, if food cannot be moved to where it is needed through trade, it may be necessary for the population to relocate. The long and heart-breaking lines of African refugees seen in recent years are one obvious example of this problem. Their destination is often a refugee camp which is secure enough for *World Food Program* or government food disbursements to take place (Kerr 2015).

The third element of food security is more long term in scope and pertains to the ability of global food production to keep pace with growing food demand arising due to a combination of expanding population and increasing incomes. If production does not increase at a pace that keeps up with rising demand, the result is a growing population chasing ever-scarcer food, which represents deterioration in food security. The potential for demand for food to outstrip the ability to increase food production—known as the Malthusian trap—has been known, since Thomas Malthus publishes his famous book *An Essay on the Principle of Population* in 1798. Thus far, Malthus' dire predictions relating to declining food security—although he did not use that term—have proved to be false. The reason for this is that he failed to take an account of the impressive increases in agricultural productivity that have occurred as a result of innovation and investments in agricultural research and development over the last two centuries. The rate at which agricultural productivity is currently increasing appears to be falling, while population continues to increase as exemplified in the projections to 2050 (Beddington 2010).

The Deputy Director of the United Nations Food and Agriculture Organization (FAO) suggests, "agricultural production needs to increase by 70% worldwide, and by almost 100% in developing countries, in order to meet growing food demand" (Tutwiler 2011). There is, however, considerable evidence of and concern regarding a serious underinvestment in agricultural research and development over the last 20–30 years (James et al. 2008; Alston et al. 2009). Even if the rate of investment could be increased, there are considerable lags between when investment expenditures are made and the resulting increases in productivity are fully manifest (Alston 2010). The lags are often in the range of 25 years or more.

The underinvestment in agricultural research has a number of causes, including long and costly regulatory processes for new technologies (Smyth et al. 2004); high costs in identifying and acquiring existing intellectual property (Smyth and Gray 2011); poor intellectual property protection (Cardwell and Kerr 2008); resistance to technological change

(Haggui et al. 2006); misaligned incentives (Malla and Gray 2005); investor's inability to capture full benefits (Alston 2002); and governments' fiscal difficulties (Gaisford et al. 2001). Research on reducing food waste exhibits an even greater degree of underinvestment. The lack of investment in agricultural research and development poses a significant food security challenge with population slated to rise considerably in the next decades.

While the degree of underinvestment in agricultural research and development is concerning, the geographic misalignment of those investments greatly increases the global food security challenge. Almost all of the increase in population and increases in income that will most affect food demand will take place in developing countries. It is these areas of the world, however, that receive the least research investment. The research capacity in developing countries is low and resource-strapped governments find it difficult to make the long-term investments required to build that capacity. As a result, if there is to be an investment in productivity-enhancing agricultural research, it will likely have to come from agribusiness firms headquartered in developed countries. These profit-driven firms, however, must see competitive returns from their investments. Thus far, they have not seen those returns and, for the most part, they eschew investments in technologies specifically designed for crops and the agronomic conditions in developing countries.[5] One of the major stumbling blocks is poor protection of intellectual property in many developing countries (Cardwell and Kerr 2008; Gaisford et al. 2007; Loppacher and Kerr 2005; Perdikis et al. 2004; Isaac and Kerr 2003). In addition, some promising technologies such as biotechnology have run into resistance in developing countries (Holtby et al. 2007) or from their trading partners, causing them to not allow the use of the technology (Smyth et al. 2013). In the face of rapidly expanding populations and incomes, if productivity improvements are inhibited in developing countries, then to maintain or enhance food security alternative sources of supply must be found.

International trade flows can provide, to a considerable degree, an alternative source of supply (Kerr 2011a). Any major gains in agricultural productivity over the next few decades are likely to arise in developed countries. Given the slower growth in population in most developed countries, much of the additional food coming from increases in productivity will be available for export. Furthermore, the major food surplus countries are concentrated in North America, around the Black Sea and in some South American countries. Basically, if there is any chance of food security being maintained or enhanced in developing countries, it will have to come from imports. Attempts at providing food security through increases in domestic production to foster self-sufficiency, as advocated by some, are bound to fail (Kerr 2011b). This does not mean that increasing local food production will not be important in meeting the challenges of food security, but it should not have as its objective food self-sufficiency. Such an objective can lead to inefficient use of resources, environmental degradation, and increased risks to food security. As suggested above, without the ability, or willingness, to trade, local crop failures can lead to a decline in food security. Given that agricultural systems in developing countries are often already stressed, pushing them harder to achieve self-sufficiency can only lead to more frequent local crop failures. The impacts of climate change are likely to exacerbate the problems associated with attempting to achieve self-sufficiency.

Barriers to international trade will inhibit the role that trade can play in enhancing food security. Protectionist-motivated trade barriers in agriculture considerably reduce the potential benefits of international trade (Gaisford and Kerr 2001). Having the potential contribution that international trade can make to food security further reduced by unintended barriers to trade is a folly and a failure in public policy. The trade problems created by asynchronous MRLs are such an unintended barrier to trade. Given the upcoming global food security challenges, such systemic unintended barriers to food trade warrant closer scrutiny to determine their source and examination of any options for removing their trade restricting elements.

Notes

1. Here we use a broad definition of production related to food availability so that production can also increase through reduction in waste as well as increases in agricultural productivity.
2. Storage of food such as saved harvest stocks can provide the means for having food available on a daily basis.
3. Or a close substitute.
4. Sometimes through their ability to hoard or, if necessary, to escape to more food secure areas.
5. Some technological advances designed for developed country crops and conditions can be transferred relatively easily to developing countries.

References

Alston, J.M. 2002. Spillovers. *The Australian Journal of Agricultural Economics* 46: 315–346.

Alston, J.M. 2010. *The Benefits from Agricultural Research and Development, Innovation, and Productivity Growth.* OECD Food, Agriculture and Fisheries Papers, No. 31. Paris: Organization for Economic Cooperation and Development Publishing. doi:10.1787/5km91nfsnkwg-en.

Alston, J.M., J.M. Beddow, and P.G. Pardey. 2009. Agricultural Research, Productivity and Food Prices in the Long Run. *Science* 325: 1009–1010.

Beddington, J. 2010. Food Security: Contributions from Science to a New and Greener Revolution. *Philosophical Transactions of the Royal Society of London—B Biological Science* 365: 61–71.

Bloy, M. 2002. The Irish Famine, 1845–9. *The Victorian Web.* http://www.victorianweb.org/history/famine.html.

Cardwell, R., and W.A. Kerr. 2008. Protecting Biotechnology IPRs in Developing Countries: Simple Analytics of a Levy Solution. *Journal of Agricultural Economics* 59 (2): 217–236.

FAO. 2008. *Briefing Paper: Hunger on the Rise—Soaring Prices Add 75 Million People to Global Hunger Rolls.* Rome: Food and Agriculture Organization of the United Nations. http://www.fao.org/Newsroom/en/news/2008/1000923/index.html.

Gaisford, J.D., J.E. Hobbs, W.A. Kerr, N. Perdikis, and M.D. Plunket. 2001. *The Economics of Biotechnology.* Cheltenham: Edward Elgar.

Gaisford, J.D., and W.A. Kerr. 2001. *Economic Analysis for International Trade Negotiations.* Cheltenham: Edward Elgar.

Gaisford, J.D., J.E. Hobbs, and W.A. Kerr. 2007. Will the TRIPS Agreement Foster Appropriate Biotechnologies for Developing Countries? *Journal of Agricultural Economics* 58 (2): 199–217.

Giordani, P., N. Rocha, and M. Ruta. 2012. *Food Prices and Multiplier Effects of Export Policy.* Geneva: World Trade Organization. http://www.wto.org/english/res_e/reser_e/ersd201208_e.pdf.

Haggui, F., P.W.B. Phillips, and R. Gray. 2006. Opposition to Genetically Modified Wheat and Global Food Security. In *International Trade and Policies for Genetically Modified Products*, ed. R.E. Evenson and V. Santaniello, 175–190. Wallingford: CABI.

Holtby, K.L., W.A. Kerr, and J.E. Hobbs. 2007. *International Environmental Liability and Barriers to Trade.* Cheltenham: Edward Elgar.

Isaac, G.E., and W.A. Kerr. 2003. International Trade, Intellectual Property Rights and Traditional Knowledge—The Case of Plant Genetic Resources. *Journal of Aboriginal Economic Development* 3 (2): 75–85.

James, J.S., P.G. Pardey, and J.M. Alston. 2008. *Agricultural R&D Policy: A Tragedy of the International Commons*. Staff Paper P08-08. Department of Applied Economics, University of Minnesota, Minneapolis. http://digitalcommons.calpoly.edu/cgi/viewcontent.cgi?article=1010&context=agb_fac&sei-redir=1&referer=http%3A%2F%2Fscholar.google.ca%2Fscholar%3Fstart%3D120%26q%3Dj%2Bm%2Balston%26hl%3Den%26as_sdt%3D0%2C5#search=%22j%20m%20alston%22.

Kerr, W.A. 2011a. The Role of International Trade in Achieving Food Security. *Journal of International Law and Trade Policy* 12 (2): 44–53. http://ageconsearch.umn.edu/bitstream/117818/2/kerr12-2.pdf.

Kerr, W.A. 2011b. Food Sovereignty—Old Protectionism in Somewhat Recycled Bottles. *African Technology Development Forum Journal* 8 (1 and 2): 4–9.

Kerr, W.A. 2015. Food Security and Anti-Piracy Strategies: The Economics of Protecting World Food Program Shipments. In *Food Security in an Uncertain World—An International Perspective*, ed. A. Schmitz, P.L. Kennedy, and T. Schmitz, 145–159. Bingly: Emerald Group Publishing.

Loppacher, L.J., and W.A. Kerr. 2005. Developing Countries' Protection of Intellectual Property Under TRIPS and Innovation Investment. *The Journal of World Intellectual Property Rights* 1 (1–2): 1–17.

Malla, S., and R. Gray. 2005. The Crowding Effects of Basic and Applied Research: A Theoretical and Empirical Analysis of an Agricultural Biotech Industry. *American Journal of Agricultural Economics* 87 (2): 423–438.

Perdikis, N., S.L. Boyd, and W.A. Kerr. 2004. Multinationals, Biotechnology, Intellectual Property and Developing Countries—Should Developing Countries Seek to Be Exploited? In *European Culture in a Changing World: Between Nationalism and Globalism*, ed. D. Meyer-Dinkgrafe, 1–10. London: Cambridge Scholars Press.

Smyth, S., P.W.B. Phillips, W.A. Kerr, and G.G. Khatchatourians. 2004. *Regulating the Liabilities of Agricultural Biotechnology*. Wallingford: CABI.

Smyth, S.J., and R. Gray. 2011. Intellectual Property Sharing Agreements in Gene Technology: Implications for Research and Commercialisation. *International Journal of Intellectual Property Management* 4 (3): 179–190.

Smyth, S.J., W.A. Kerr, and P.W.B. Phillips. 2013. Accelerating the Adoption of Genetically Modified Crops in Africa Through a Trade Liability Regime. *Plant Biotechnology Journal* 11: 527–534.

Tutwiler, A. 2011. Abstract for Keynote Address to the GMCC-11 Conference, October 26–28, Vancouver, Canada.

World Bank. 2009. *Global Economic Prospects 2009: Commodity Markets at the Crossroads*. Washington: World Bank. http://siteresources.worldbank.org/INTGEP2009/Resources/10363_WebPDF-w47.pdf.

CHAPTER 3

Incentives, or Not, for Governments to Remove Asynchronous MRLs as Trade Barriers

Abstract While there are good economic and food security reasons to pursue international harmonization of MRLs, there are constraints on how it can be achieved. Both policy makers and those charged with implementing harmonization face incentives that do not align with the need to eliminate asynchronous MRL regulatory regimes across countries. Progress in harmonization of MRLs has, as a result, proved to be elusive.

Keywords International trade · Non-tariff barriers · Disincentives · Consumer confidence · Incentives · Standards · Domestic political factors Regulators · Officials · Agencies · Risk communication

While a multitude of trade barriers persists, nearly all countries recognize the need for a global trading system that is fair, transparent, and well functioning for both imported and exported agri-food products. The creation of the World Trade Organization (WTO) during the Uruguay Round of General Agreement on Tariffs and Trade negotiations in the mid-1990s, along with agreement on a multitude of new disciplines that member countries have voluntarily agreed to abide by, is evidence of that belief. By creating the WTO and the disciplines within it, countries recognized that global trade was growing rapidly and becoming increasingly complex and diverse. There was also a recognition that trade

M.T. Yeung et al., *Declining International Cooperation on Pesticide Regulation*, Palgrave Studies in Agricultural Economics and Food Policy, DOI 10.1007/978-3-319-60552-4_3

liberalization was not solely about reducing tariffs, but also ensuring that non-tariff barriers (NTB) do not unnecessarily impact the flow of trade.

Trade in agricultural commodities and food products has always been, and probably always will be, a complicated and sensitive subject (Gaisford and Kerr 2001). A simultaneous argument could be made that international movements of agri-food products are a vital component of international trade. Governments depend on it for generating wealth through exports, countries are provided with a varied, stable, and sustainable food supply through imports, and consumers have access to lower cost, wider choice, and greater availability of food.

The creation of the WTO was a significant step forward in addressing the NTBs in agricultural trade. It established numerous provisions to facilitate trade and reduced the ability of countries to implement unjustified domestic measures intended to protect their domestic agricultural sectors. There are four parts of the WTO agreement targeting NTBs in agri-food products. The first is the Agreement on the Application of Sanitary and Phyto-sanitary Measures (SPS) which outlines what a country can and cannot do with respect to food safety and animal or plant health. The recognition of three previously existing science-based institutions: (1) Codex; (2) the World Organization for Animal Health (OIE); and (3) the International Plant Protection Convention (IPPC) as the international standard setting bodies for food and agriculture products are the second. The third is the Agreement on Technical Barriers to Trade (TBT) which outlines what a country can and cannot do with regard to other technical standards that are not related to food safety and animal or plant health. Finally, the WTO includes a dispute-settlement mechanism that provides a means to ensure compliance with the rules and disciplines.

While many NTBs exist, the global asynchronicity of MRLs is a significant and growing impediment to the international trade of agricultural commodities and food products. The literature review of Chap. 5 and the case studies in Chap. 6 provide several examples where non-harmonized MRLs have disrupted trade resulting in increased cost for industry and governments, and impaired the international flow of agri-food products.

The question arises: if Codex, the international body responsible establishing food safety standards through a rigorous scientific process, is setting MRLs, why then is there a problem with MRLs and why are governments not simply adopting the international standards? There are several disincentives for governments to do that.

First, pressure on governments to maintain domestic consumer confidence in the food supply has increased dramatically since the mid-1990s. Food safety scares and easy access to (mis)information, combined with increasingly vocal, non-government organizations critical of the use of chemicals in the production of food and the technologies and practices used to ensure safe usage, have all served to erode consumer confidence. Consumer skepticism regarding the regulatory systems responsible for the safety of their food is rising globally, resulting in increasing demand for governments to take measures that better protect their domestic food supplies, whether produced at home or abroad (Nakuja et al. 2015; Liu et al. 2009).

While those governments that have the capacity to do so generally do a good job at risk assessment and risk mitigation, their risk communication to consumers continues to present a significant challenge. By way of example, it is likely that the overwhelming majority of consumers worldwide have never heard of Codex, do not know what it does, or how its work relates to their own individual well-being. It is understandable then, that when seeking confidence in their food supply, consumers are not looking towards an organization based in Rome that they have not heard of. They are looking, instead, towards their own governments to develop a *made-at-home* standard that they can trust.

In order to gain, regain, or maintain the trust of their citizens regarding food safety, many governments have resorted to developing their own MRLs, sometimes but not consistently referencing or basing them on Codex standards, but nevertheless creating a domestic MRL. Furthermore, governments often enact in domestic legislation requirements that go beyond the mandate and work of Codex, and, therefore, leave them with little or no legal maneuverability to adopt international MRLs.

Governments also need to fill the gaps when an international standard does not exist. There are a number of reasons why Codex is unable to develop standards when they are needed and those reasons will be elaborated upon later in this volume. Hence, in the absence of an international MRL, or a domestic MRL, and in the face of increased consumer scrutiny, governments are less and less able to accept the exporting country's MRL or a regional MRL. As a result, governments sometimes resort to a default tolerance of zero, which, in effect, often makes the continuation of international trade impossible (Hobbs et al. 2014).

There are many reasons why national governments feel the need to act on their own. There are, however, several compelling incentives to collaborate globally to address NTBs. Perhaps, the most important is ensuring that, as a global community, we continue to allow trade in agri-food products to contribute to the achievement of food security goals, and to mitigate regulatory barriers that impede achieving them. As suggested previously, by 2050, the global population will exceed nine billion people, but the increase in arable land for crop production will increase by only 4%. While new agricultural technologies will contribute productivity induced growth in the food supply, there will be an even greater need to avoid creating an overly complicated and unworkable regulatory system that could seriously hinder trade and the movement of agricultural commodities and food products.

While this will be important to all countries, it will be of particular importance to developing and least-developed countries dependent on consistent and stable agri-food imports to feed their growing populations, as well as for generating wealth for their producers as some of them will increasingly become exporters of food products. For countries which are large exporters and increasingly looking to new markets in the developing and least-developed countries (with their growing populations and rising incomes), there is a large vested interest in ensuring that those markets have the capacity to accept imports.

The challenge will be in resisting the current trend of countries *going-it-on-their-own* to develop import standards and move again towards working in international fora to arrive at mutually beneficial, common solutions. For MRLs, this means that there should be a significant incentive for all countries to strengthen Codex and ensure that it has the scientific capacity and resources to develop standards acceptable for both importers and exporters.

Another incentive for countries to work collaboratively in international bodies is the more efficient use of scarce time and resources in developing standards. Maintaining the human, knowledge, and physical resources necessary to develop standards is costly, so it begs the question as to why a country would invest those resources to develop a MRL when an international one may already exist. Perhaps, those resources would be better spent somewhere else in the domestic food safety system.

Finally, one cannot ignore the fact that domestic political factors always impact agricultural trade. The crux of most trade negotiations is what concessions countries are willing to make in order to liberalize the market. With reductions in tariffs, NTBs are the next natural tool of choice for protectionist interests seeking to shelter their producers and agri-food industries from international competition (Gaisford and Kerr 2001). It is important to continue challenging measures that are not based on science and that unjustifiably impede trade (Smyth et al. 2013). It is critical that all nations work together bilaterally, regionally, and multilaterally through the WTO, Codex, and all other available avenues to ensure that political factors do not mitigate the importance of having a global trading system that provides certainty for consumers, producers, processors, importers, and exporters in aid of enhancing food security.

As suggested above, it has been agreed multilaterally that the global trade of agri-food products must be consistently based on sound science to provide those engaged in exporting and importing with a reliable, predictable, and stable environment for engaging in international commerce (Isaac 2007). However, bringing together trade policy and scientific regulation can be challenging. Regulators are primarily concerned with ensuring that the science is thorough, rigorous, accurate, and justifiable compared to the mandate of trade officials, who often have to address a more complicated set of issues when managing trade. Those issues can be socially, civically, or politically driven, including factors such as economic priorities, industry goals, consumer preferences, availability of governmental resources, and inter-agency rivalries, to name but a few.

The WTO SPS Agreement provides the rules on how science and trade policy should work together; however, it does not specifically direct countries as to how they achieve that balance. Furthermore, it does not obligate countries to provide the science on which a measure has been based, but it does encourage countries to be transparent through notification to the WTO. The SPS also gives countries a fair amount of flexibility to put in restrictions on imports for a vague *reasonable period of time* under Article 5.7 which deals with incomplete scientific information (Smyth et al. 2009).

Hence, while countries have certain WTO obligations pertaining to science and trade, they also have a great deal of flexibility in setting MRLs and the science upon which they are based or, in some cases, not based upon. Further complicating the matter is that within each country, several agencies can be involved in establishing a MRL—a standard-setting agency, an international trade department, an inspection agency, an agriculture department, health and environment agencies, and so on—making it extremely difficult for trading partners to obtain a complete understanding of how a MRL is being established and how to have input in or influence its development.

The presence of this broad group of agencies and departments, each with its own specific mandates, can lead to tensions amongst the players. As an example, when trade officials would like a certain standard in place to facilitate trade flows, and regulators want to maintain the integrity of the regulatory process and not be influenced by trade goals, tensions often occur. When these tensions do arise, and with the array of agencies and departments that may be involved in the standard-setting process, it can be extremely difficult for an outsider to identify who is in the best position to provide a solution and who is influencing the outcome. Such inter-agency tension can slow or even preclude regulatory reform, for example, in the establishment and/or implementation of a standard as negotiated in a trade agreement.

Unfortunately, there is no magic bullet to deal with this situation. It is a complex matter of building networks, strengthening relationships, and opening communication channels. The front line to do this is, generally, government diplomats abroad in embassies and missions, with trade officials working with their trade counterparts, as well as scientists and technical experts, if they happen to be posted abroad, working with the regulators who are doing the scientific work and developing the MRL. Domestically based trade officials and regulators also have several opportunities for communicative outreach to foreign counterparts as well as through in-person meetings at international fora such as Codex, the WTO, Asia Pacific Economic Cooperation (APEC), and scientific conferences to name a few.

Commodity groups and crop protection developers also have an important role to play both domestically, with their governments and by working within international industry associations. The establishment of clear goals across the value chain or across sectors and then communicating those goals to governments is critical. Building alliances with foreign industry counterparts and like-minded industries and countries must happen in order to inform the work of government trade officials and regulators.

All of the above has focused on issues and responses for working with foreign governments bilaterally as they move away from international

MRLs and increasingly move to toward *made-at-home* domestic ones. Responding on a *one-by-one* basis to individual countries' MRLs is not efficient or desirable. If anything remotely resembles the *magic bullet* that prevents this, it is to move countries back to Codex and encourage the adoption international standards. As Codex is a joint body of the World Health Organization (WHO) and FAO, the scientific undertakings of the Joint Meetings on Pesticide Residues (JMPR) and the Codex Committee on Pesticide Residues (CCPR) to arrive at the adoption of a MRL need to respond to both health and safety, as well as commerce. Codex also provides a forum for those countries that want to participate in the establishment of a MRL, to do so, whether it is to have input into the science, debate the data, influence the conclusion, or raise objections if necessary.

How do we get there? The first important step, whether dealing with a specific country or encouraging the development and adoption of Codex MRLs, is for governments and industry to agree on a strategy to move forward—a written detailed strategy—that includes elements such as the problem to be addressed; the objective; the government bodies that need to be influenced; benchmarks; the messages to carry forward; who is best messenger; and where and when to deliver those messages and so on.

The next challenge relates to how does the industry facing asynchronous MRLs in its export markets, and/or conversely, an importer trying to obtain agri-food products from international suppliers, encourage its own government agencies to undertake the complex issue of some form of MRL harmonization.

References

Gaisford, J.D., and W.A. Kerr. 2001. *Economic Analysis for International Trade Negotiations.* Cheltenham: Edward Elgar.

Hobbs, J.E., W.A. Kerr, and S.J. Smyth. 2014. How Low Can You Go? The Consequences of Zero Tolerance. *AgBioForum* 16 (3): 207–221.

Isaac, G.E. 2007. Sanitary and Phytosanitary Issues. In *Handbook on International Trade Policy*, ed. W.A. Kerr and J.D. Gaisford, 383–393. Cheltenham: Edward Elgar.

Liu, H., W.A. Kerr, and J.E. Hobbs. 2009. Product Safety, Collateral Damage and Trade Policy Responses: Restoring Confidence in China's Exports. *Journal of World Trade* 43 (1): 97–124.

Nakuja, T., M. Akhand, J.E. Hobbs, and W.A. Kerr. 2015. Evolving US Food Safety Regulations and International Competitors: Implementation Dynamics. *International Journal on Food System Dynamics* 6 (4): 259–268.

Smyth, S.J., P.W.B. Phillips, and W.A. Kerr. 2009. Global Governance Quandaries Regarding Transformative Technologies for Bioproducts, Crops and Foods. *Journal of World Trade* 43 (6): 1299–1323.

Smyth, S.J., W.A. Kerr, and P.W.B. Philips. 2013. Managing Trade in Products of Biotechnology—Which Alternative to Choose: Science or Politics? *AgBioForum* 16 (2): 126–139.

CHAPTER 4

The Economics of International Harmonization of MRLs

Abstract The economics of asynchronous regulations internationally is modeled. The model is used to show the economic costs associated with the persistence of regulatory divergence. Given the economic and food security benefits from international regulatory harmonization, factors that enter into decision-making that are not captured in economic assessments are chronicled.

Keywords Economic model · Transactions · Competitiveness · Harmonization · Equivalence · Switching costs · Regulatory convergence · Welfare · Import ban · Regulatory divergence · Costs · Type one errors · Bureaucracy · Science-based · Minimum threshold · Negotiations · Political precaution · Risk

4.1 The Problem

Differing MRLs for pesticides among countries is part of a wider trade issue pertaining to regulatory standards that can act as barriers to international trade. Mandatory standards represent government regulatory requirements that must be met by private sector economic actors (firms or individuals) before they are allowed to enter into transactions in markets under the jurisdiction of the particular government which has established the requirements. This rule applies to both domestic firms and foreign firms that wish to enter into transactions in that particular

M.T. Yeung et al., *Declining International Cooperation on Pesticide Regulation*, Palgrave Studies in Agricultural Economics and Food Policy, DOI 10.1007/978-3-319-60552-4_4

regulated market. When the regulatory requirements that a firm must satisfy in order to undertake transactions in its home market differ from those of the importing markets the firm wishes to export to, then that firm may not be able to satisfy the regulatory requirements of the foreign market, meaning that it cannot enter into transactions in the foreign market—it cannot engage in international trade—or it can only enter into transactions in the exporting market after incurring additional costs—its international competitiveness is reduced, thus inhibiting its ability to trade. As it is these regulatory differences that lead to the reduction in international trade, eliminating those regulatory differences will increase international trade.

The differences in regulatory regimes that disrupt trade can be removed in several ways, but two are the most compelling: (1) regulatory harmonization among countries; (2) the granting of equivalence for foreign regulations. Regulatory harmonization requires that the countries involved devise and agree on a common set of regulations that will apply in their respective jurisdictions. With regulatory differences removed, international trade will no longer be fettered by regulatory divergence. In the case of two countries, A and B, there are three possible paths to harmonization; (1) country B agrees to adjust its regulatory regime and adopts (harmonizes to) the regulatory regime of country A; (2) country A agrees to adjust its regulatory regime and adopts (harmonizes to) the regulatory regime of country B; (3) both countries agree to adjust their current regulatory regimes and to devise a new common regulatory regime.

Adjusting to a different regulatory regime will require firms and governments to incur costs—switching costs. In the first case above, all of the switching costs will be borne by firms and the government in country B and none by country A. In the second case, all of the switching costs will be borne by firms and the government in country A and none by country B, while in the latter case, both country A and country B will have to bear some of the switching costs. Thus, countries have an incentive to negotiate hard to ensure that their existing standards become the common (harmonized) standard. Ongoing monitoring by both countries is prudent to ensure that its trading partner is enforcing the common regulatory regime with sufficient diligence. If regulations are harmonized, desired changes to regulations may be delayed by the need to negotiate the change with trading partners.

In the case of the granting of equivalence, each country agrees to recognize the other county's regulatory standard as satisfying its own regulatory requirements. In other words, while the requirements may differ, it is perceived that the efficacy of the foreign standard is at least equal (or sufficiently close to equal) to the efficacy of the domestic standard. If the foreign regulatory regime is not judged to be as effective as the domestic standard by an unacceptable degree then equivalence should not be granted.[1]

If equivalence is granted, the barrier to trade will be removed and there will not be any switching costs. Ongoing monitoring is, again, prudent to ensure that trading partners are enforcing their standards sufficiently to ensure that equivalence is maintained. If a country decides to alter its regulatory regime, then the trading partner needs to evaluate the new regulatory regime to ensure that equivalence is maintained.

4.2 The Economics of Regulatory Convergence

If harmonization or the granting of equivalence is achieved, the removal of the trade barrier will be welfare enhancing. In the case of harmonization, it will be welfare enhancing if the gain in the welfare in the market is larger than the switching cost. This can be illustrated in Figs. 4.1 and 4.2.

In Fig. 4.1, an exporter can supply the importer's market at P_1. Assume that there is however, a divergence in regulations between the importing country and the exporting country. If the divergence is in a form whereby the exporter cannot satisfy the importer's requirements, then neither imports nor exports can occur. This is equivalent to an import ban (Gaisford and Kerr 2001). The price in the importer's market will be P_0 and the quantity produced and consumed in the market is Q_0. Assume now that the divergence between the regulatory regimes could be removed, either through the harmonization of the regulatory regimes or the granting of equivalence by the importing country. As a result of the removal of the barrier to trade, imports can now take place. Price in the market will fall to P_1 and imports of Q_{1d} minus Q_{1s} will take place. The domestic supply and demand functions in the importing country remain unchanged (where the intersection of the s and d lines represent domestic equilibrium) as does the ability of the exporter to supply the market at P_1—in other words, the efficacy of the regulatory regime is unchanged and costs are not altered for either the exporter or import competing firms.

As a result of the removal of regulatory divergence, the total welfare derived from the market increases from an area with a monetary value

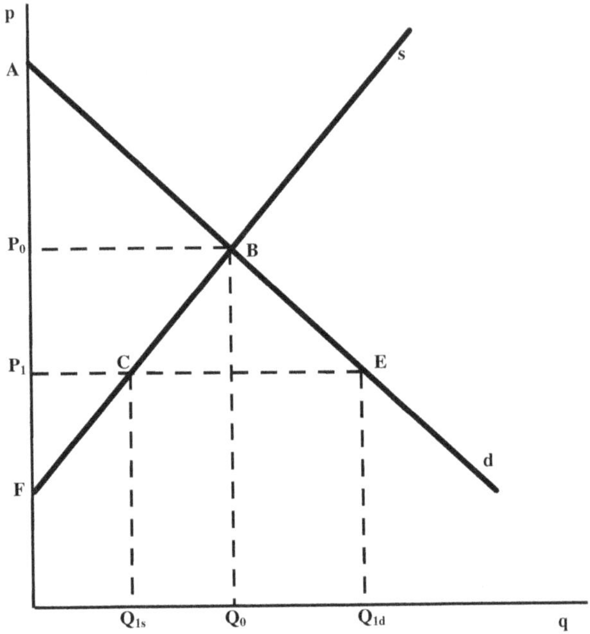

Fig. 4.1 Regulatory divergence acts as an import ban

equal to ABF to an area with a monetary value equal to ACEF—or a net increase in welfare equal in value to BEC. Exporters can make sales in the market equal in value to $CEQ_{1d}Q_{1s}$. The removal of regulatory divergence is, thus, unambiguously welfare increasing for society.

In Fig. 4.2, regulatory divergence imposes additional costs on exporters through the extra costs incurred to comply with the importer's regulatory regime. Unfettered, the exporter can supply the importer's market at P_1. This form of the regulatory divergence means, however, that to satisfy the importer's regulatory requirements and to be able to have its product enter the importer's market, the exporter faces increased costs equal to P_0 minus P_1. Thus, the price in the importer's market rises to P_0. At P_0, imports equal Q_{0d} minus Q_{0s}. The value received by the exporter for its exports is equal to area $VUQ_{0s}Q_{0d}$ with area YXUV lost in additional regulatory costs. This is equivalent to the imposition of a tariff on imports (Gaisford and Kerr 2001).

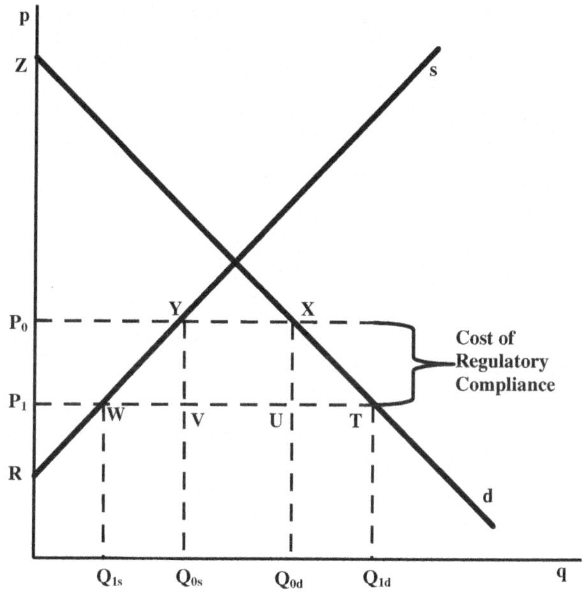

Fig. 4.2 Regulatory divergence raises compliance costs for exporters

If the regulatory divergence can be removed, either through harmonization or the granting of equivalence by the importer, then the additional cost will no longer exist and the market price will fall from P_0 to P_1. At P_1, imports increase from Q_{0d} minus Q_{0s} to Q_{1d} minus Q_{1s}. The total welfare arising from the market increases from ZXYR to ZTWR, or an unambiguous increase in total welfare of YXTW. The exporter has an increase in the value of sales of $(WVQ_{0s}Q_{1s})$ plus $(UTQ_{1d}Q_{0d})$.

Both of these results are standard in the trade literature (Gaisford and Kerr 2001). The actual size of the values will depend on the individual market under consideration. In the case of harmonization, if there is a new common regulatory regime, there may be increases or decreases in the domestic supply function, the export supply function or both, depending on the new costs associated with compliance. If these cost changes were sufficiently large to eliminate any gain in total welfare, then regulatory convergence should not be undertaken.

In both cases, the important point is that there is an unambiguous gain in society's welfare.

Thus, policy makers interested in the welfare of society should grant equivalence, where there is no adjustment cost, or engage in a process of harmonization if the value of the increase in welfare is greater than the adjustment cost (BEC > adjustment cost in Fig. 4.1 or YXTW > adjustment cost in Fig. 4.2).

Of course, the figures above are a simplistic representation of markets which could be subject to a process of regulatory convergence. For example, they depict one time period, whereas the benefits can be expected to continue over a number of time periods. Evaluations normally consider this through a process that computes the benefits in future time periods and then discounts future benefits. This does not alter the essential result that regulatory convergence leads to net positive benefits under normal situations (Kerr 2006a).

Given the unambiguous welfare enhancing results arising from economic theory, one should expect to see regulatory convergence. When it does not take place, it leads to considerable frustration among those who can see economic opportunities from the removal of regulatory divergence and those concerned with global food security. It also implies that explanations for the difficulties in realizing regulatory convergence lie outside the welfare maximizing motivations provided by economic theory (Kerr 1997, 2006b).

4.3 Factors Inhibiting Regulatory Convergence Omitted from Economic Evaluations

If economic evaluations often suggest that regulatory convergence is unambiguously welfare enhancing, the question naturally arises as to why it (often) does not come to pass. It suggests that there must be other factors at work that inhibit public policy initiatives to foster regulatory convergence. There may be two broad reasons for this. First, some factors may simply be too difficult to incorporate in any economic assessments undertaken. The best economic analysis will attempt to include such factors, but the assumptions that often have to be made due to incomplete or absent information mean that the credibility of the analysis will not improve. This is well known and, as a result, they are often left out of the analysis. Policy makers may be well aware of such factors when they are omitted from formal analysis and will notionally value them in their decisions. Second, there may be other, non-economic factors, that impact policy-makers' decisions—something else is considered more important than welfare maximization. Ultimately, policy-makers have to make decisions and information on expected increases in welfare is only one factor in the decision-making process. Other factors may include:

4.3.1 Type One Errors

One of the factors that are difficult to formally include in economic assessments is the probability of making a *type one error* and the consequences of making a *type one error*. No decision can be error-free. This is well recognized in the decision-making literature. Two types of errors have been identified (Van den Belt 2003):

> *Type one error*: This is where a positive decision is made—say to harmonize MRLs—when it should not have been made, i.e., they should not have been harmonized.
>
> *Type two error*: This is when a negative decision is made—say not to harmonize MRLs—when a positive decision should have been made, i.e., they should have been.

In the case of *Type two errors*, the discussion is about the opportunities forgone due to having made a decision error. Opportunities forgone in the future are, however, often difficult to assess and if they do not come to pass, the beneficiaries may simply remain ignorant of the benefits they have missed. In an unknown future, it is difficult to see what might have been.

In the case of *Type one errors*, the consequences are more often obvious and can be extremely costly. One well-known example is the approval of the drug *thalidomide* for use by pregnant women. It was a *Type one error* which had horrific consequences for those directly affected by taking the drug. It also had serious consequences for those who made the *error* and indirectly for all those involved in the approval processes for pharmaceuticals. While errors cannot be eliminated, policy-makers will factor the potential to make a *Type one error* into their decision processes, and may take actions to reduce the probability of making a *Type one error* in spite of any net welfare benefits that may apparently be forgone due to a *Type two error*.

The probability of making a *Type one error* is unknown to decision-makers. In the case of a regulatory convergence issue, the decision-maker cannot know whether the probability of making a *Type one error* is increased by either granting equivalence to foreign regulations or by harmonizing—unless the other party harmonizes to their country's regulatory regime. Given that foreign procedures or alternatives arrived at through harmonization are likely less well understood, a risk-averse policy-maker may opt for maintaining the divergent *status quo*.

Furthermore, maintaining national control of regulatory processes may make it easier to deal with the consequences of a *Type one error*. Another example of a *Type one error* was the European Union's decision to treat *Bovine Spongiform Encephalopathy* (BSE)—also known as *mad cow disease*—as solely a livestock disease (Kerr et al. 2007; Loader and Hobbs 1996). As a result, while infected animals were dealt with, no provisions were made to protect meat intended for human food supply chains. When it was subsequently determined that a potential link between BSE and the human disease *variant Creutzfeldt-Jakob* (vCJD) existed, it was clear that a *Type one error* had been made. The human disease is fatal and no cure exists. A number of processes were quickly put in place to eliminate meat of potentially infected animals from human food supply chains and a much more aggressive program targeted at eradicating the animal disease was implemented. The major outbreak was in the United Kingdom (UK), although some cases were found in other parts of the EU. Beyond the individuals that contracted vCJD from consuming meat from BSE-infected animals, there was considerable fallout from this *Type one error*. The credibility of the EU, and particularly the UK, food safety systems, was damaged (Loader and Hobbs 1996), requiring major changes. Arguably, it was a major contributor to the establishment of a new food safety regime in the EU—the European Food Safety Authority (EFSA). Furthermore, it led to the perception that the British Ministry of Agriculture was captured by farming interests. It also led to the end of one political career, the UK Minister of Agriculture at the time who publically insisted beef was safe and attempted to forcefully make the point by feeding his daughter a beef burger at a public event where the media were present. His political career did not survive this *Type one error*.

The problems associated with having to respond to the BSE crisis, however, had wider ramifications. The EU has a *single market*, meaning that there is a significant degree of regulatory convergence, largely accomplished through harmonization. The UK was a major exporter of beef to the EU. At the height of the BSE crisis, all UK exports to the continent were halted. Once the UK government had dealt with the BSE crisis in ways which complied with EU-wide standards, it wanted beef exports to resume. Some individual governments in the EU did not want to allow the return of British beef into their market, fearing another *Type one error*. They had, however, moved to harmonized EU-wide standards under the *single market* and, hence, had no mechanism to legally deny market access to British beef. Despite this lack of legal mechanism, market access for British beef continued to be denied, resulting in long and acrimonious disputes between the European Commission and the UK government. It would have been much easier for both the British government and the governments of individual member states of the EU to deal with the consequences of the *Type one error* pertaining to BSE if harmonization had never taken place. This is a lesson that has not been lost on regulators, not just in the EU but globally (Kerr et al. 2007; Loppacher and Kerr 2005). Even if the probability of making a *Type one error* is extremely low, given the bias toward caution built into international norms such as the Risk Assessment Framework that is used to assess new technologies (Isaac 2007), the potential for such errors cannot be entirely eliminated. It is difficult to assess the weight decision-makers that put on the problem of *Type one errors* and little is known about how such weightings can be reduced.

4.3.2 *Incentives and the Bureaucracy*

Initiatives related to regulatory convergence are often under the ambit of the ministry responsible for international trade. On the other hand, unlike for example, the removal of tariffs, implementation of regulatory convergence agreed to in trade agreements is often under the mandate of ministries whose primary responsibility is not trade liberalization (Kerr 1997). Their primary responsibility is to ensure the health and safety of domestic consumers or the natural environment. They may regard foreign standards, or even new harmonized standards, with suspicion. They have little incentive to alter their *tried and true* systems despite what officials in the trade ministry may have agreed to. Basically, Ministry of Health officials or those in food inspection agencies have no mandate to consider benefits from trade and are unlikely to be swayed by arguments pointing out the welfare gains that could be made from reducing trade barriers.

Despite the perception that social science is 'soft' and does not provide definitive answers and that, alternatively, science is precise and definitive, the reality is quite different. Science is far from a monolith. While the broad outlines of the scientific method may be widely accepted, there is considerable scope for divergence in implementation, and in particular when devising regulations pertaining to human, animal, or plant health or the environment. If science were monolithic, there would not be regulatory divergence among countries. These differences among science-based regulatory regimes arise for a number of reasons. Scientists in different countries are trained in different ways. Over time procedures can develop in different ways and become embedded in the scientific education and the scientific culture of a country. These differences can then become part of the science-based regulatory process. Thus, one should not expect that regulatory convergence should be a natural process—and it may be fiercely resisted.

This is because in every country, those charged with ensuring health and safety will be doing the best job that they can—at least in their own minds. If they were not, then they would not be doing their job. This means that there is an inherent bias towards believing that their current existing system must be superior to all other existing systems. If it is not, then those charged with ensuring health and safety should improve the system. To admit that a foreign system is superior (or sometimes even equivalent) to the system currently in use is to admit that one is not doing one's job. If harmonization of regulations—either to an existing foreign standard or a newly negotiated harmonized standard—could lead to a superior outcome, then why was not it implemented unilaterally? Thus, admitting that any alternative standards are superior can lead to awkward questions which can have wider ramifications for those in the existing bureaucracy. This fosters a belief in the superiority of domestic mechanisms and a reason to resist changes outlined in trade agreements. It can explain why a little progress is made when commitments are made to discuss harmonization of standards in trade agreements (Kerr 1992).

This can also explain the slow progress at institutions charged with harmonization such as the Codex, the OIE, and the IPPC. These are institutions where, for the most part, science-based regulators attempt harmonization—in other words, they are to a large degree divorced from the political realm. If each country's representatives perceive that their current set of regulatory procedures as superior to all others, and that any change will lead to a reduction in the health and safety outcomes for those within their jurisdictions, one should not be surprised that progress is slow. To admit the methods of others might be superior or to agree to alternative standards both imply that regulators are not doing their jobs. Such admissions can have consequences that damage the regulatory system's credibility and can have a negative impact on individual careers.

Harmonization is likely to require changes to existing procedures which may simply be resisted due to a dislike of change among some members of the bureaucracy—it may mean changes to staffing levels or diminished responsibility that can adversely affect career prospects. Furthermore, changes may require additional resources for everything from alternative scientific instruments to retraining. Such budgetary increases may simply not be forthcoming from ministries or legislatures (Nakuja et al. 2015). Hence, if implementation goes forward, the bureaucracy will be faced with doing *more with less*, which may be a popular political slogan in some quarters but hardly a positive incentive for change within the bureaucracy.

In some cases, the changes that arise from harmonization may actually lead to the loss of jobs within the bureaucracy if certain procedures are dropped or skills no longer deemed relevant (Kerr 1988, 1997). Individuals who might be adversely affected in this way have a clear incentive to resist regulatory convergence.

4.3.3 Limitations on Negotiating Resources

Often, despite expressed ambitions to remove regulatory divergence through trade agreements, little progress is actually made. Some of this is because science-based regulations are inherently technical and beyond the comfort zone of trade negotiators. Thus, trade agreements can only yield venues to consider/negotiate technical issues. In trade agreements, however, while venues are established, they seldom have closure mechanisms to ensure that negotiations will actually be completed (Kerr and Hobbs 2015; Kerr 2016). In other words, these venues are only forums to *talk and talk* (Kerr 2006b).

Those with trade negotiation skills are a scarce resource. Trade negotiations are very wide ranging and time-consuming, forcing trade negotiators to prioritize. The goal of trade negotiators is to obtain an agreement that is the best deal possible once some *minimum threshold* is achieved. If the *minimum threshold* cannot be achieved, there will be no agreement. Regulatory convergence is seldom a core part of the *minimum threshold.* Thus, negotiators tend to focus first on achieving the objectives constituting the *minimum threshold.* Once that is achieved, then, achieving objectives on a *wish list* can be attempted. Negotiators interested in achieving the best agreement possible must assess the probability of success in achieving the various items on the wish list. Those which have a low probability of success are abandoned, or *kicked down the road* to be achieved in future negotiations. Given the inherent resistance to regulatory divergence discussed above, the probability of achieving such an objective is likely low. It is also relatively easy, given the requirement for technical discussions, to reach agreement on establishing a venue to *talk and talk,* which has the appearance of progress toward regulatory convergence when in fact it is not. An example is the establishment of a venue to discuss the regulation of biotechnology in the Comprehensive Economic and Trade Agreement between the EU and Canada (Kerr and Hobbs 2015). A number of similar institutional arrangements were embedded in the Canada–US Trade Agreement/ North American Free Trade Agreement which did not yield regulatory convergence (Smyth et al. 2015).

4.3.4 Political Precaution

Beyond bureaucratic resistance to regulatory convergence, considerations beyond welfare maximization may also become of concern at the political level. Events or circumstances negatively impacting the ability to govern or to continue to govern in the future are of interest to politicians. The negative effects of events may impact the broader exercise of governance or individual politicians. The impacts may be as small as having to answer difficult questions to as large as having to resign. The impacts of events have been amplified in recent years with the emergence of social media which can increase both the spread of negative perceptions and the speed at which the information is disseminated. Politicians may become aware of potential or actual problems either from constituents, lobbyists, or the media. Politicians may either wish to act proactively to prevent an event with a potentially negative impact or to be reactive to limit the potential damage. This often results in a precautionary approach to policies or events.

In recent years, *political precaution* has become a phenomenon of politics, and in particular when issues of human health and safety are involved (Kerr 2009). *Political precaution* has been defined as arising:

> when politicians are being pressured to "do something, or to be "seen to be doing something" in the face of strongly expressed concerns by members of civil society even when risks are very low or largely speculative (Kerr 2004, p. 40).

Isaac (2007) identifies three types of risks: (1) recognized risks (for which there are data for determining risk); (2) hypothetical risks (for which there are no data but acceptable analytical methods for determining risks) and (3):

> speculative risks, which lack experience, data, a causal-consequence mechanism and an accepted analytical method for assessment; they are logical possibilities—irrefutable and untestable. Typically, such risks would have no standing within a science-based framework (p. 288).

Members of civil society may be clamoring for politicians to act on the basis of speculative risks—a pesticide residual may cause cancer—when there is no evidence to suggest a link between the pesticide and cancer. Politicians may wish to act to counter such speculative threats to reduce the clamoring for action and/or to reduce the potential damage to their ability to govern. Examples include import bans on pork by China and a number of other countries during the H1N1 pandemic in spite of there being no link between consuming pork and the H1N1 virus (Kerr 2009), the EU import ban on beef produced using growth hormones (Kerr and Hobbs 2002), and EU restrictions on imports of the products of modern agricultural biotechnology (Isaac and Kerr 2007; Viju et al. 2012).

Once an issue becomes *politicized* and *political precaution* becomes the motivation for the decisions of politicians, it is unlikely that sovereignty will be conceded. Thus, regulatory convergence will be resisted and regulatory independence, with its inherent flexibility, preferred by politicians over more constraining alternatives. Arguments pertaining to the economic benefits associated with regulatory convergence will not carry much weight once an issue becomes politicized.

4.3.5 Economic Protection

Removal of trade barriers, either through harmonization or the granting of equivalence, will lead to economic losers as well as winners. Those who are potential losers have an incentive to ask politicians for protection. This is the normal reason for ***protectionist*** requests. If regulatory divergence provides protection, then regulatory convergence can be resisted through lobbying for protection. The potential losers can be seen in Figs. 4.1 and 4.2 above. In the case of regulatory divergence acting as an import ban, removal of the barrier will lead to a loss in producers' surplus in the importing country—area P_0BCP_1 in Fig. 4.1; and area P_0YWP_1 in Fig. 4.2 when the regulatory barrier is cost increasing. The welfare gain from liberalization will always be larger than the loss suffered by protectionist vested interests. The granting of economic protection is a political decision and the arguments surrounding it and how it may be controlled as old as the study of economics—it was one of the issues dealt with by Adam Smith in his *Wealth of Nations* in 1776.

4.4 Achieving International Regulatory Convergence for MRLs Will not Be Easy

While regulatory convergence likely leads to an increase in economic welfare, regulatory divergence persists. This is because other considerations factor into decisions pertaining to regulatory convergence. If these factors carry more weight with decision-makers than the expected economic benefit, then continued regulatory divergence is likely to persist, often to the considerable frustration of those who can see economic gains, both for their own enterprise and society in general.

Leaving aside the problem of *Type one errors*, scarcity of negotiating resources and requests for economic protection, the potential for achieving regulatory convergence can be summarized in Table 4.1.

Table 4.1 Resistance to regulatory convergence

		Politicized issue	
		Yes	**No**
Pre-existing regulations	**Yes**	Bureaucratic resistance ***and*** political resistance	Bureaucratic resistance ***but*** no political resistance
	No	No bureaucratic resistance ***but*** political resistance	No bureaucratic resistance ***and*** no political resistance

When there are an already existing set of domestic regulations, one is likely to encounter bureaucratic resistance to regulatory convergence. On the other hand, if regulations are to be developed where none exist, the bureaucracy has less reason to resist. If the issue has not become politicized, then there is less likelihood of resistance from politicians. If, however, the issue has become politicized, then political resistance to regulatory convergence can be expected.

Clearly, it will be most difficult to achieve regulatory convergence when there are both pre-existing domestic regulations and the issue has become politicized—the top left-hand box in Table 4.1. On the other hand, the best chance for achieving regulatory convergence is when no pre-existing regulations are in place and the issue is not politicized—the bottom right box in Table 4.1. Different strategies will have to be applied in the case where only bureaucratic resistance can be expected—the top-right-hand box in Table 4.1 or where the issue is politicized, but no bureaucratic resistance can be expected—bottom-right-hand box in Table 4.1.

Note

1. See Kerr (2004) for a discussion of when granting of equivalence may not be warranted.

References

Gaisford, J.D., and W.A. Kerr. 2001. *Economic Analysis for International Trade Negotiations*. Cheltenham: Edward Elgar.

Isaac, G.E. 2007. Sanitary and Phytosanitary Issues. In *Handbook on International Trade Policy*, ed. W.A. Kerr, and J.D. Gaisford, 383–393. Cheltenham: Edward Elgar.

Isaac, G.E., and W.A. Kerr. 2007. Whose Vision of the Future? The Entrenched International Conflict Over Genetic Modification. *The Geneva Post Quarterly* 2 (1): 87–107.

Kerr, W.A. 1988. The Canada-United States Free Trade Agreement and the Livestock Sector: The Second Stage Negotiations. *Canadian Journal of Agricultural Economics* 36 (4): 895–903.

Kerr, W.A. 1992. Removing Nontariff Barriers to Trade Under the Canada-United States Trade Agreement: The Case for Reciprocal Beef Grading. *Journal of Agricultural Taxation and Law* 14 (3): 273–288.

Kerr, W.A. 1997. Removing Health, Sanitary and Technical Non-Tariff Barriers in NAFTA. *Journal of World Trade* 31 (5): 57–73.

Kerr, W.A. 2004. Sanitary Barriers and International Trade Governance Issues for the NAFTA Beef Market. In *Keeping the Borders Open, Proceedings of the Eighth Agricultural and Food Policy Systems Information Workshop*, ed. R.M.A. Loyns, K. Meilke, R.D. Knutson, and A. Yunez-Naude, 26–49. Guelph: University of Guelph.

Kerr, W.A. 2006a. International Harmonization and the Gains from Trade. *Journal of International Law and Trade Policy* 7 (2): 116–125. http://ageconsearch.umn.edu/bitstream/23843/1/07020116.pdf.

Kerr, W.A. 2006b. NAFTA's Underdeveloped Institutions: Did They Contribute to the BSE Crisis? In *Agrifood Regulatory and Policy Integration Under Stress*, ed. K.M. Huff, K.D. Meilke, R.D. Knutson, R.F. Ochoa and J. Rud, 213–226. Guelph: Texas A and M University, University of Guelph, Instituto Interamericano de Cooperacion para la Agricultural-Mexico.

Kerr, W.A. 2009. Political Precaution, Pandemics and Protectionism. *Journal of International Law and Trade Policy* 10 (2): 1–14. http://ageconsearch.umn.edu/bitstream/52231/2/kerr10-2.pdf.

Kerr, W.A., and J.E. Hobbs. 2002. The North American-European Union Dispute Over Beef Produced Using Growth Hormones: A Major Test for the New International Trade Regime. *The World Economy* 25 (2): 283–296.

Kerr, W.A., and J.E. Hobbs. 2015. A Protectionist Bargain? Agriculture in the European Union-Canada Trade Agreement. *Journal of World Trade* 49 (3): 437–456.

Kerr, W.A., L.J. Loppacher, and J.E. Hobbs. 2007. International Standards for Regulating Trade When BSE is Present: Why are They Being Ignored? *Current Agriculture, Food and Resource Issues* 8: 1–15.

Kerr, W.A. 2016. Worlds Apart on GMOs—Can Trade Agreements Bridge the GAP? NABC Report 27. In*Proceedings of the 27th North American Agricultural Biotechnology Council Conference*, Pennsylvania State University, 133–145, June 2–3, 2015, North American Agricultural Biotechnology Council, Ithaca.

Loader, R., and J.E. Hobbs. 1996. The Hidden Costs and Benefits of BSE. *British Food Journal* 98 (11): 25–34.

Loppacher, L.J., and W.A. Kerr. 2005. The Efficacy of World Trade Organization Rules on Sanitary Barriers: Bovine Spongiform Encephalopathy in North America. *Journal of World Trade* 39 (3): 427–443.

Nakuja, T., M. Akhand, J.E. Hobbs, and W.A. Kerr. 2015. Evolving US Food Safety Regulations and International Competitors: Implementation Dynamics. *International Journal on Food System Dynamics* 6 (4): 259–268.

Smyth, S.J., P.W.B. Phillips, and W.A. Kerr. 2015. Food Security and the Evaluation of Risk. *Global Food Security* 4: 16–23.

Van den Belt, H. 2003. Debating the Precautionary Principle: "Guilty Until Proven Innocent" or "Innocent Until Proven Guilty"? *Plant Physiology* 132: 1122–1126.

Viju, C., M.T. Yeung, and W.A. Kerr. 2012. The Trade Implications of the Post-Moratorium European Union Approval System for Genetically Modified Organisms. *Journal of World Trade* 46 (5): 1207–1238.

CHAPTER 5

Previous Examinations of MRLs

Abstract A review of the literature on MRLs and their effect on international trade is presented. The literature is far from extensive but, in general, confirms the propositions regarding the effects of asynchronous regulatory regimes among countries. Reduced international trade levels are reported. The nature of the observed barriers to trade arising from asynchronous MRL regulations is explained.

Keywords Asynchronous regulatory regimes · Standards misalignment Econometric modeling · Distribution · Quantification · Stringency · Developing countries · Least developed countries · Adjustment costs · Welfare enhancing · Consumers · Technical barriers to trade · Literature review · Effects · Gravity model

The issue of international convergence of MRLs has not received a great deal of attention in academic work. Those studies pertaining to MRL misalignment have focussed on a limited range of issues including:

1. Attempts to quantify the impact of MRL (or food safety standards) misalignment, for specific groups, sectors, or global trade flows using econometric modelling;

M.T. Yeung et al., *Declining International Cooperation on Pesticide Regulation*, Palgrave Studies in Agricultural Economics and Food Policy, DOI 10.1007/978-3-319-60552-4_5

2. Analysis of the distribution of economic consequences of MRL misalignment on trade flows and in particular, between developed and developing countries as well as other stakeholders;
3. Advising that econometric quantification of estimates of impacts should be treated carefully due to shortcomings in the ability to include or quantify exogenous factors;
4. Suggesting that MRLs cannot be treated solely as benefiting or detracting from trade as the distribution of effects is not spread evenly among or between countries;
5. Assessments of the challenges surrounding harmonization of MRLs;
6. Identifying that domestic MRL stringency can influence the ability to obtain market access when exporting;
7. Indicating that MRL harmonization is welfare enhancing, even after adjustment costs are taken into account;
8. Finding that MRL misalignment persists even when positive economic outcomes are identified;
9. Identifying that consumers, particularly those with higher incomes, are increasingly demanding increased food safety standards as well as increased stringency in MRLs.

Several studies address standards or technical barriers to trade rather than specifically focusing on MRLs; however, the analysis and impacts are applicable to MRLs. There are a number of studies that provided important insights regarding MRL harmonization.

Smyth et al. (2015) warn that the increasingly piecemeal implementation of barriers to commodity trade can be countered with bilateral and multilateral agreements to adopt harmonized regulatory processes, resulting in economic gains. Despite harmonization being agreed to in a trade agreement, however, the actual implementation of harmonization requires a commitment of resources. Reasons for inertia against harmonization include conflicting priorities between government agencies and trade negotiators, vested interests in preserving the status quo due to risk aversion, methodological egocentrism, and bureaucratic resistance to change.

Swinnen and Vandemoortele (2011) find that consumers are increasingly demanding standards, because they reduce the risks associated with information asymmetry, particularly when applied to health or quality issues of a product. Their model measures the effect that political weight has on a politically optimal standard, taking into account the costs faced by domestic and foreign producers, the domestic consumer demanding the standard and the impact that lobbying has on administrative agencies overseeing the standard. They note that such standards can act as both barriers and catalysts to trade.

Winter and Jara (2015) show that MRLs have been diverted from their original intent as signalling the use of GAP in the production of food to that of a direct indicator of food safety by those worried about risks (often speculative) pertaining to consumption. They propose a new approach, using pesticide food safety standards (PFSS) to develop scientifically defensible levels regarding safety concern. The PFSS approach is demonstrated through the example of the use of the fungicide captan on strawberries. Their results show that PFSS levels are many times greater than the existing tolerance and MRL levels, scientifically proving that detected residue levels that are significantly higher than MRLs do not pose acute or chronic health concerns, even though the MRL level may be violated.

Wilson and Otsuki (2003) illustrate how divergent international food safety standards can impact global agri-food trade flows using the example of aflatoxin standards. A gravity model is used to estimate the impact that different aflatoxin standards in 15 (4 developed) importing countries have on the exports of 31 countries (21 developing). Results indicate that divergent standards have the greatest negative effect on the exports of least developed countries. Their model estimates that if the international Codex standard was adopted, world exports would increase by $38.8 billion.

Handford et al. (2015) provide a summary of global pesticide legislation, including a discussion on the development of pesticide legislation and divergences across countries. Noting that the EU standardized its members' individual MRLs to a single European one, the EU maintains more stringent MRLs for most pesticides compared to the US, which has the least stringent MRLs in some cases. Exporters defer to the EU or Codex MRLs, though divergence still impedes trade.

Handford et al. (2015) also find that exporting producers already meeting stringent domestic MRLs at home are more capable of complying with stringent MRLs of destination nations. MRL stringency places a larger burden upon developing countries for a number of reasons—pesticides not authorized in developed country markets are often used; a lack of institutional and infrastructure resources in pesticide regulation including awareness of risks, labeling, storage, transport, disposal, and underdeveloped registration systems and; issues with compliance and enforcement. A common practice is to increase monitoring, with the potential to ban products if a non-compliant situation persists from that source country.

They note that divergent pesticide legislation and standards are driven by a number of motivations including culture, politics, economics, science (differing test standards, methodology, and measurements), health, food sustainability and security, safety, trade, pest management, and state of development. Pesticide use varies by geography, needs, application, and climate. Divergence is the result. There are also different GAP for different application rates depending on the pest being treated, causing further variance. The absence of a MRL can also impede trade.

Several challenges to harmonization exist. These include high transactions costs—switching costs—in changing existing systems and regulations being unique to a country's own domestic needs, which cannot be addressed by a single global standard. Harmonization is also limited by individual countries' domestic policies, goals, and priorities which, first, determine the resources that they will allocate to the issue and, second, may conflict and compete with international goals and priorities. Also hindering harmonization is scientists presenting varying interpretations, methodologies, and opinions; developed countries have their own strategic priorities and agendas; developing countries lack resources to undertake pesticide reviews on their own. Handford et al. (2015) suggest that efforts to move towards harmonization should include the improvement of Codex procedures; consistent use of the OECD tools; and support of the work of the OECD towards harmonizing pesticide registration systems such as a consistent OECD format for all pesticide submissions. Global reviews conducted via work sharing and collaboration amongst developing countries to assist that their regulatory programs are also recommended.

They believe that successful efforts at harmonization will require research programs to address knowledge gaps regarding pesticides risks; individual countries need to rank by importance, the key drivers of pesticide regulation to determine which drivers to prioritize; current regulatory systems should be evaluated for their strengths and weaknesses to assist in finding a system that can act as a common denominator, but one that can also accommodate individual countries' needs; assessment made of the benefits and risks of a harmonized system for all concerned parties; and improvements should be made to the knowledge base of concerned parties, so that legislation is fully implemented and enforced.

Keener et al. (2014) note the US Food and Drug Administration (FDA)'s estimate that US consumers purchased $2 trillion worth of imported products from 8,25,000 importers through more than 300 US ports of entry in 2007—and the volume of imports could double every 5 years. They describe various US FDA strategies to work with global agencies and organizations toward regulatory capacity-building, harmonization, awareness-building, and information sharing to facilitate responses to public health emergencies. They recommend that the best strategy to address food safety regulation is to be proactive and take on issues before they become politicized. As all matters involving national autonomy such as legislation and regulation are political, achieving consensus on what constitutes a safe food supply will be inherently political. The interpretation and implementation of SPS agreements are increasingly controversial and emotionally charged in the politics of international trade. Examples include the issues of domestic security and public health such as the US bioterrorism laws and the European Union's food and feed regulations.

Xiong and Beghin (2012) caution against black and white interpretations of MRLs acting purely as either barriers to, or facilitators of, trade. As examples of where MRLs which deviate from Codex standards act as barriers to trade, they cite:

1. Fisher and Serra (2000)—a social planner maximizing domestic welfare sets standards that are too stringent compared to those of a global social planner seeking to maximize global welfare (domestic welfare and foreign firms' profit).
2. Marette and Beghin (2010)—a policy-maker would adopt excessively stringent standards to correct externalities when its domestic producers have a comparative advantage in complying with the regulations but would do the opposite if their producers are much less efficient than foreign competitors in meeting the standards.
3. Berti and Falvey (2011) who assess the interface between trade and stringency of standards in a political economy model.

As examples where food safety standards that address information asymmetry or consumption externalities can lead to increased demand, they cite:

1. Jaffee and Henson (2005) who show that after successfully upgrading the supply chain to meet EU sanitary standards, Kenyan

fresh vegetable growers were able to expand their exports to the EU.

2. Xiong and Beghin (2011) who indicated that intra-OECD agricultural trade is facilitated by technical measures, because the regulations enhance the import demand more than that they penalize foreign suppliers meeting this import demand through incurring higher costs.

Xiong and Beghin (2012) caution that the use of models or any other empirical measurement and quantification of the stringency of standards is difficult to undertake as are the determination of globally optimal standards. Many factors cannot be accounted for in the models.

They then illustrate how food safety standards affect imports as well as domestic competitiveness and export potential by using a gravity equation approach to trade, with empirical analysis of pesticides and veterinary drug MRLs. They compare the US and Canada. Their results show that for the US, which is systematically equivalent to Codex, imports from rest of world are not impeded, but US exports are constrained by destination market MRL regulations. Compared to the US, Canada's stricter pesticide MRLs (with average MRLs being well above Codex MRLs in several sectors), have facilitated Canadian international competitiveness without excluding foreign competition from its domestic market. This is likely for two reasons. First, because they already meet strict domestic regulations, Canadian exporters find meeting strict MRL requirements of other nations less challenging. Second, lower tolerance levels (i.e., stricter standards) in Canada could enhance Canadian exporters' reputation in foreign markets where food safety pays off in terms of either a higher premium or repeated purchases.

Xiong and Beghin (2012) conclude that the export performance of producers complying with stringent domestic standards at home can either promote exports or hinder them; it is a function of whether those producers have a comparative advantage in complying with the regulations and to what extent the enhanced safety or improved quality is appreciated by consumers in destination markets. They cite:

1. Li and Beghin (2012) indicate, specifically pertaining to MRLs on pesticide and veterinary drug residues, countries adopting stringency in excess of Codex, which tend to be developed ones, benefit their own exports.

2. Farnsworth (2012) reports that the levels of MRLs are attributable to public concerns about food safety and countries' development status, but less to industry's lobbying efforts, although no formal structural model is developed to rationalize these empirical findings.
3. Li (2012) finds that MRL stringency and tariffs are jointly determined, which suggests that political decision-making might have a role in the design or implementation of MRL policies.

Xiong and Beghin's (2012) analysis pertaining to the potential harmonization in NTB measures between the US and Canada suggests that Canada may be less inclined to ***harmonize downward*** by relaxing its stringency to US or Codex levels due to loss of international competitiveness. On the other hand, the US would be reluctant to ***harmonize upward*** to Canadian MRLs, because setting standards systematically above Codex might unnecessarily compromise the domestic industry's competitiveness and risk losing the reputation of a free trader in the world market.

Disdier et al. (2007) used the gravity approach and data on the use of SPS and TBT notifications to the WTO to assess impact on trade flows and distribution of impact among countries. They found that the majority of agri-food products face some sort of barrier in importing countries and that, generally, SPS and TBT measures negatively impact imports of OECD countries. Their estimates suggest that while SPS and TBT barriers do not significantly affect the bilateral trade between OECD members, they do reduce developing and least developed countries' exports to OECD countries considerably. Within Cairns Group[1] and non-Cairns members', exports are similarly affected by SPS and TBT regulations. Furthermore, EU imports seem to be more negatively influenced by SPS and TBTs than imports of other OECD countries. Finally, analysis by sector shows that SPS and TBT measures could foster trade in some sectors.

Farnsworth (2012) uses ordered logit and probit regression analyses to estimate whether MRL stringency is being utilized as a protectionist trade barrier. The results indicate that rich, food importing countries employing fewer people in agriculture that also spend a higher percent of their GDP on public health, favor stricter regulations. Patterns of protectionism at the commodity and country level were not found, though isolated examples may exist. The possible logical explanation given is that regulators successfully erect MRLs to reflect the government's priorities regarding health while still maintaining a reasonable level of harmonization. If economic protectionism is present, it is neither obvious nor widespread.

Countries with high total imports in agriculture, higher consumer incomes, and public health spending and/or higher health awareness tend to have stricter MRLs. Those with greater employment in agriculture and those that produce the regulated commodities subject to MRLs tend to have looser MRLs.

Martinez and Thornsbury (2010) note that divergent MRLs create an additional barrier to trade for small and mid-sized firms, in particular, as access to information in a readily available and understandable format is a challenge. The consequences of a border rejection are often bans on future trade and/or increasingly stringent inspection requirements (and associated costs) for that particular exporter or the entire product category, thus increasing costs industry-wide. They also indicate that specialty crops are at a particular disadvantage in a system of lowest detectable levels as they often involve minor uses for many compounds and, hence, are often overlooked and/or regulated by default categories that may not directly apply.

They cite several studies illustrating the impact that MRL regulations can have on fruit:

1. Wilson and Otsuki (2004) found that stringent regulation of the pesticide chloropyrifos by OECD countries on banana exports leads to a decrease of imports of 63%.
2. Otsuki et al. (2001) showed that stringent MRL regulation in the UK caused a 64% reduction in the exports of cereals from African countries. They also estimated an approximate US$ 5.5 billion in lost banana exports and US$670 million in lost cereal exports per year when international standards were set at the EU levels of regulatory stringency instead of using the Codex standard.
3. Dinham (2003) provides an example where, in the 1990s, vegetable exports from Guatemala destined for the US were rejected due to high residue levels and resulted in a decline in Guatemalan farm prices of more than 40%; this was an entirely preventable collapse.

Winchester et al. (2012) estimate the impact of non-harmonized trade regulations by using a heterogeneity index of trade regulations (HIT) specific to bilateral trade flows and a variant that accounts for whether heterogeneity is due to more stringent regulations in the importing country. The study examines the trade flows of the EU and nine of its partners and also measured the impact of NTBs on bilateral trade in plant products. Their analysis uses HIT sub-indexes for import requirements in a gravity model to examine the trade impact of differences in regulations across countries while accounting for tariff barriers. Of note, their study does not account for when an exporter complies with stricter MRLs at home than in the destination market.

One of their findings is that stricter pesticide MRLs for plant products in one country, relative to other countries', reduce exports to that initial country. This suggests that regulators should invest in efforts to ameliorate divergence in pesticide MRLs across countries, for example, by adopting internationally agreed MRLs like those suggested by the Codex. Their results also showed that, on the whole, despite utilizing several indexes, regulatory heterogeneity does not have a negative and significant impact on trade, although their sample covered only a small number of plant products. They suggest that differences are not wide enough to act as trade impediments.

Beghin et al. (2015) study the effect of NTBs and standards in trade and global value chains. They note that the complexity and nuances of economic and political economy effects complicate any attempts at empirical measurements and their use in policy. Their literature review cites:

1. Baldwin (1970) where for a social planner addressing an externality in an open economy context with domestic and foreign firms competing for the domestic market, the optimum policy to address the external effect should maximize world/global welfare.
2. Marette and Beghin (2010) where the optimum policy to maximize domestic welfare (as opposed to global welfare) is not necessarily protectionist as it depends on the relative ability of domestic and foreign industries to comply with the corrective policy.
3. Swinnen and Vandemoortele (2008, 2009, 2011), where the lobbying of domestic firms and consumers may lead to standards being set *too low* or *too high* in the political equilibrium, depending on among other things, the relative costs of compliance.

In assessing whether a standard-like NTB is protectionist, Beghin et al. (2015) suggest several criteria:

a. If the measure is not addressing any market imperfection based on scientific evidence and a risk assessment (external effect, asymmetric information leading to health or environment hazard), then the policy is protectionist, except if it is set for legitimate precautionary reasons. The latter condition means that while science is being established, a restriction or policy impediment can be put in place temporarily.
b. National treatment also provides useful guidance as differences in standard-like NTBs imposed on like-products (domestic versus imported) would signal protectionism (Baldwin 2000).
c. Third, transparency has to prevail and trade partners have to be notified of, and consulted about, the policy in a predictable manner. The WTO also calls for least-trade restrictive policies. This last criterion is more controversial, because maximizing trade opportunities while addressing the external effect may not maximize welfare (Disdier and Marette 2010). For example, the WTO does not account explicitly for the impact of standards on consumer welfare through remaining market imperfections.

They suggest a caveat that econometric models using gravity equations to measure bilateral trade flows should be interpreted with care as often the models do not provide domestic or international welfare impact estimations. The models also account for decreasing tariffs as generalized commodities rather than specific products. Finally, Beghin et al. (2015) recommend that policy making bodies should increase compatibility of NTBs across trading nations via reciprocity, equivalence, or mutual recognition, and certifications of international standards.

Shortall (2007) examines the work undertaken by the OECD to identify regulatory tools, processes, and policies that support market openness and improve trade. Pertaining to regulations, the OECD recommends using international standards, mutual recognition agreements, cooperative arrangements, equivalency, inter-departmental/agency domestic regulatory coordination, and national and subnational cooperation and coordination.

The example of Canada is given where a less formal approach to interdepartmental coordination is used via its Trade and Regulatory Interdepartmental Committee. The committee, which is chaired by the head of the delegation to the TBT Committee, brings together representatives of regulatory departments and agencies, the office responsible for regulatory oversight, and the Standards Council of Canada. The committee meets prior to each TBT Committee meeting and provides ongoing policy advice on regulatory and trade issues.

Shortall (2007) notes that countries resort to different mechanisms to ensure consultation, coordination, and cooperation between regulatory authorities at different levels of government. An example of a formal committee structure designed to address common policy issues through consultation and information exchange is Canada's Federal/Provincial/Territorial Working Group on Regulatory Governance. Created in 2005, it provides a forum to lay the foundation for common principles on managing regulation, with a focus to establish common regulatory principles, institute a common approach to regulatory impact assessment, and share best practices.

Drogue and Demaria (2013) conduct an empirical analysis on 40 countries exporting and/or importing apples, pears, and processed products. Their results show that overall, regulatory heterogeneity may be trade-impeding and that international harmonization of the applicable MRL might have a positive impact on trade. A case-by-case analysis, however, showed that adopting a single regulation would not have a positive impact for all countries. Harmonization would even have a negative impact on exports of pears and apples from Japan and the US. Their study does not account for exporters complying with stricter MRL regulations in their home market than in their destination markets.

Xiong and Beghin (2014) show that technical regulations can facilitate production and increase demand by enhancing plant, animal, and human health. By examining MRLs on the import demand of OECD countries, they found that MRLs jointly enhanced import demand by increasing consumer awareness of food safety thus stimulating demand for products under regulation. However, MRLs hindered non-compliant export supply and exporters from LDCs were more constrained by MRLs than their developed world competitors.

Foletti and Shingal (2014) study the effects of heterogeneity in MRL pesticide regulation on bilateral trade by assembling panel data for 50 countries between 2006–2012, assessing 118 products and 1193 pesticides. They found that regulatory heterogeneity both increased and deceased trade. Their results show trade increased for exports coming from countries that set the strictest standards, alluding to the positive informative effect of such regulations. Trade decreased when exporters faced more stringent regulation abroad, suggesting compliance costs in entering destination markets. They also found that technical regulations can deliberately or unwittingly impede trade. They specifically included the effect on exports when the exporting country maintains more stringent MRL regulations at home than in destination markets.

Finally, Chen et al. (2015) discuss the institutional and administrative bodies and processes established in China to manage pesticides, including the supervisory process and system, research, and standards setting. MRLs are established on a residual chemical assessment, a toxicity assessment, and a dietary intake assessment. Median residue and the highest residue amounts, acceptable daily intake, and acute reference dose are determined through testing and are then combined with the dietary intake assessment and the nutrition and health status of the Chinese people to obtain a recommended MRL for pesticides.

They note that China's recent mandatory pesticide standard (GB 2763–2014) was published in 2014; it provides 3650 MRL standards for 387 kinds of pesticides in 284 types of foods, with coverage over the majority of commonly used pesticides and foods most often consumed. MRL standards for primary processing products (such as fruit juices, preserved fruits, and dried fruits) are included. The new standard also focused on establishing pesticide MRLs in fresh foods, especially in vegetables and fruits. In total, 2495 MRL standards of pesticides were established for 115 kinds of vegetables and 85 kinds of fruits. The new standard also uses 1999 Codex Alimentarius Commission MRL Standards; 90.6% of China's MRLs are set more stringently than or are equal to the Codex levels.

Chen et al. (2015) describe China's activities at the Codex Committee on Pesticide Residues as well as its progress in constructing a pesticide management system. A standard system for pesticide residues has been improved, evaluation procedures for MRLs have been established, and coordination with Codex on pesticide residues has, for the most part, been achieved. China has also increased its capacity to be involved in the establishment of the international standards for pesticide residues. Six MRLs standards from China have been adopted by Codex (endosulfan and cypermethrin in tea, and acephate and its metabolite methamidophos in brown rice and straw). China is also actively pursuing and expanding international exchange and cooperation especially with the FAO, the EU and the US in the fields of standard establishment, detection technology, and risk assessment.

The results of the review of literature on regulatory divergence are summarized in Table 5.1.

Table 5.1 Summary of common findings in academic literature relevant to MRLs

Non-harmonized food safety standards, including MRLs, are most detrimental to developing and least developed countries	Wilson and Otsuki (2003); Handford et al. (2015); Disdier et al. (2007); Xiong and Beghin (2014)
Exporters whose home countries maintain stringent domestic MRLs have less difficulty in complying with foreign market MRLs as they are already compliant	Handford et al. (2015); Xiong and Beghin (2012); Foletti and Shingal (2014)
Effects of divergence in MRLs can increase and decrease trade, for both developed and developing countries. Effects are not equally distributed	Xiong and Beghin(2012, 2014); Winchester et al. (2012); Disdier (2007); Foletti and Shingal (2014), Swinnen and Vandemoortele (2011); Drogue and Demaria (2013)
Political issues affect the ability to harmonize MRL policy in ways that consider factors other than the economic benefits	Handford et al. (2015); Keener et al. (2014); Beghin et al. (2015); Swinnen and Vandemoortele (2011); Smyth et al. (2015); Xiong and Beghin (2012)
There is a likely correlation between consumers' (whether domestic or foreign) awareness of health and food safety and increasing demand for products subject to stringent MRLs	Xiong and Beghin (2012); Farnsworth (2012); Xiong and Beghin (2014)
Despite uneven distribution of benefits and incurring adjustment costs, working towards some form of harmonization is likely to be net welfare enhancing	Smyth et al. (2015); Wilson and Otsuki (2003); Winchester et al. (2012); Drogue and Demaria (2013)

While studies of the trade effects of MRLs and other NTBs provide insights into both the extent and complexity of the problem, additional understanding of the issues can be gleaned from case studies. A selection of case studies pertaining to trade problems and MRLs is presented in the next section.

Note

1. The Cairns Group is a coalition of agricultural exporting nations committed to reforming agricultural trade. Members of the group are Argentina, Australia, Brazil, Canada, Chile, Colombia, Costa Rica, Guatemala, Indonesia, Malaysia, New Zealand, Pakistan, Paraguay, Peru, Philippines, South Africa, Thailand, Uruguay, and Vietnam.

References

Baldwin, R.E. 1970. *Nontariff Distortions of International Trade*. Washington: The Brookings Institution.

Baldwin, R.E. 2000. Regulatory Protectionism, Developing Nations, and a Two-Tier World Trade System. *Brookings Trade Forum*. Washington: Brookings Institution Press, 237–293.

Beghin, J., M. Maertens, and J. Swinnen. 2015. Non-Tariff Measures and Standards in Trade and Global Value Chains. Working Paper 15-WP 554, Center for Agricultural and Rural Development, Iowa State University, Ames.

Berti, K. and R. Falvey. 2011. Does Trade Weaken Product Quality Standards? Research Paper 2011/24, University of Nottingham Research Paper Series, University of Nottingham, Nottingham.

Chen, Z., F. Dong, J. Xu, X. Liu, and Y. Zheng. 2015. Management of Pesticide Residues in China. *Journal of Integrative Agriculture* 14 (11): 2319–2327.

Dinham, B. 2003. Growing Vegetables in Developing Countries for Local Urban Populations and Export Markets: Problems Confronting Small-scale Producers. *Pest Management Science* 59 (5): 575–582.

Disdier, A.C., and S. Marette. 2010. The Combination of Gravity and Welfare Approaches for Evaluating Nontariff Measures. *American Journal of Agricultural Economics* 92 (3): 713–726.

Disdier, A.C., L. Fontagné, and M. Mimouni. 2007. The Impact of Regulations on Agricultural Trade: Evidence from SPS and TBT Agreements. CEPII Working Paper No 2007–04, Centre d'Études Prospectives et d'Informations Internationales: Paris.

Drogue, S. and F. Demaria. 2013. Apples, Pears and Pesticides Impact of Heterogeneous Regulations Governing Pesticide Residues on World Trade, INRA No. 3, Institut National de la Recherche Agronomique: Paris.

Farnsworth, D. 2012. MRLs: Protectionism or Food Safety? Paper Prepared for Presentation at the Agricultural and Applied Economics Association 2012 Annual Meeting, Aug 12–14, Seattle.

Fisher, R., and P. Serra. 2000. Standards and Protection—Are They Protectionist. *Journal of International Economics* 52 (2): 377–400.

Foletti, L., and A. Shingal. 2014. *Stricter Regulation Boosts Exports: The Case of Maximum Residue Levels in Pesticides*. MPRA Paper No. 59895, University Library of Munich, Munich.

Handford, C., C. Elliott, and K. Campbell. 2015. A Review of the Global Pesticide Legislation and the Scale of Challenge in Reaching the Global Harmonization of Food Safety Standards. *Journal of Integrated Environmental Assessment and Management* 11 (4): 525–536.

Jaffee, S., and S. Henson. 2005. Agro-food Exports from Developing Countries: The Challenges Posed by Standards, Global Agricultural Trade and Developing Countries. In *Global Agricultural Trade and Developing Countries*, ed. M.A. Aksoy and J.C. Beghi, 91–114. Washington: The World Bank.

Keener, L., S. Nicholson-Keener, and T. Koutchma. 2014. Harmonization of Legislation and Regulations to Achieve Food Safety: US and Canada Perspective. *Journal of the Science of Food and Agriculture* 94 (10): 1947–1953.

Li, Y. 2012. *The Political Economy of Food Standard Determination—International Evidence from MRLs*. Ames: Mimeo, Iowa State University.

Li, Y., and J.C. Beghin. 2012. Protectionism Indices for Non-Tariff Measures: An Application to Maximum Residue Levels. Working Paper No. 12013, Department of Economics, Iowa State University, Ames.

Marette, S., and J.C. Beghin. 2010. Are Standards Always Protectionist? *Review of International Economics* 18 (1): 179–192.

Martinez, L., and S. Thornsbury. 2010. Identifying Maximum Residue Limit (MRL) Regulations Faced by Michigan Fruit Industries, Ag Econ Report #640, Department of Agricultural, Food and Resource Economics, Michigan State University, East Lansing

Otsuki, T., J.S. Wilson, and M. Sewadeh. 2001. Saving Two in a Billion: Quantifying the Trade Effect of European Food Safety Standards on African Exports. *Food Policy* 26 (5): 495–514.

Shortall, D. 2007. Regulatory Reform and Market Openness: Processes to Assess Effectively the Trade and Investment Impact of Regulation. OECD Trade Policy Papers, No. 48, Organization for Economic Development and Cooperation, Paris.

Smyth, S.J., P.W.B. Phillips, and W.A. Kerr. 2015. Food Security and the Evaluation of Risk. *Global Food Security* 4: 16–23.

Swinnen, J.F.M., and T. Vandemoortele. 2008. The Political Economy of Nutrition and Health Standards in Food Markets. *Applied Economic Perspectives and Policy* 30 (3): 460–468.

Swinnen, J.F.M., and T. Vandemoortele. 2009. Are Food Safety Standards Different from Other Food Standards? A Political Economy Perspective. *European Review of Agricultural Economics* 36 (4): 507–523.

Swinnen, J.F.M., and T. Vandemoortele. 2011. Trade and Political Economy of Food Standards. *Journal of Agricultural Economics* 62 (2): 259–280.

Wilson, J.S., and T. Otsuki. 2003. Food Safety and Trade: Winners and Losers in a Non-Harmonized World. *Journal of Economic Integration* 18 (2): 266–287.

Wilson, J.S., and T. Otsuki. 2004. To Spray or Not to Spray: Pesticides, Banana Exports, and Food Safety. *Food Policy* 29 (2): 131–146.

Winchester, N., M.L. Rau, C. Goetz, B. Larue, T. Otsuki, K. Shutes, C. Wieck, H. Burnquist, M. Pinto de Souza, and R. Nunes de Faria. 2012. The Impact of Regulatory Heterogeneity on Agrifood Trade. *The World Economy* 35 (8): 973–993.

Winter, C., and E. Jara. 2015. Pesticide Food Safety Standards as Companions to Tolerances and Maximum Residue Limits. *Journal of Integrative Agriculture* 14 (11): 2358–2364.

Xiong, B., and J. Beghin. 2014. Disentangling Demand Enhancing and Trade Cost Effects of Maximum Residue Regulations. *Economic Inquiry* 52 (3): 1190–1203.

Xiong, B., and J. Beghin. 2012. Stringent Maximum Residue Limits, Protectionism, and Competitiveness: The Cases of the US and Canada. IATRC Working Paper, International Agricultural Trade Research Consortium, Washington.

Xiong, B., and J.C. Beghin. 2011. Disentangling the Demand-Enhancing Effect and the Trade-Cost Effect of Technical Measures Among OECD Countries. Working Paper No. 11019, Department of Economics, Iowa State University, Ames.

CHAPTER 6

Case Studies of Trade Problems Related to MRLs

Abstract A number of case studies that provide deeper insights into the effects of asynchronous regulatory regimes for MRLs are presented. Cases include: Philippine banana exports to South Korea; imports of tree nuts into the EU; cocoa exports from Ghana; North American cranberries destined for the EU; cherry imports into Taiwan; and global avocado trade, among others.

Keywords Regime revision · Generic products · WTO notification · Extension · Default tolerance · Zero tolerance · Violations · Import tolerance · Authorized · Methodology · Inspection · Testing · Unauthorized substance · Rejection · Positive list

The following cases are illustrative examples of how misaligned MRLs are affecting international trade flows and are provided to demonstrate concrete examples of how MRLs are creating significant and costly problems for those engaged in international commerce. This list is a sample and by no means exhaustive. In every case below, the product was compliant with domestic regulations and posed no health risk, but encountered significant challenges due to MRLs.

M.T. Yeung et al., *Declining International Cooperation on Pesticide Regulation*, Palgrave Studies in Agricultural Economics and Food Policy, DOI 10.1007/978-3-319-60552-4_6

6.1 Philippine Bananas and South Korea

South Korea's imports of bananas are valued at, on average, US$250 million annually. The Philippines is the primary source, supplying 97% of this demand. In 2014, South Korean banana imports were valued at over US$320 million, with the Philippines supplying just over US$305 million of this. Bananas are the most popular fruit among South Korean consumers, comprising 48% of the country's overall fruit imports in 2012. Bananas have enjoyed this status amongst South Korean consumers for more than a decade.

As part of a major revision of its overall MRL regime, South Korea reduced the MRL for the fungicide iprodione, from 5 to 0.02 mg/kg at the end of August 2014. This new reduced MRL is, for example, significantly less than the Japanese MRL for Philippine bananas which is set 10 mg/kg, and also more stringent than South Korea's own domestic iprodione MRL for apples and pears (10 mg/kg). A senior official with the Filipino Banana Growers and Exporters Association stated to local media that the South Korean measure affected popularity of the fruit, and reduced its saleability by as much as 20% between August and September 2014.

The Philippine banana industry was not contesting the new MRL, but only the short timeline provided for adjustment. The WTO was notified of the new reduced MRL, but opposition from the Philippines was not registered. South Korea plans to progressively lower MRLs on other pesticides as well.

The new reduced MRL seriously challenges the viability of Philippine banana exports to South Korea. The Philippines was seeking an extension of the existing MRL into 2018; if this is not achieved, South Korean consumers could face either shortages of their preferred fruit, or sharp increases in the price for what a little supply is available. Suppliers from alternative banana exporting countries are facing the same MRL reduction, and/or will not be able to replace such a large market share in a short period of time.

In December 2016, South Korea's Ministry of Food and Drugs Safety (MFDS) ruled in favour of several MRLs requested by the Philippines for pesticides used in the production of bananas. Rather than defaulting to the zero tolerance MRL of 0.01 ppm if a MRL was missing, MFDS granted import tolerances for several pesticides used by the Philippines in banana production that have been disallowed in domestic use. Due to the favorable MRL decisions, Filipino banana producers can continue to access the South Korean market.

This scenario will repeat for other commodities as South Korea moves forward in implementing its own positive list MRL system, effective in January 2017 for tropical fruits, nuts, and seeds, and for additional products in December 2018.

The sources for this case study are: Diega (2015), Arcalas (2016), Business Mirror (2015), AsiaFruit (2013), and USDA (2012; n.d.).

6.2 US Tree Nuts and the EU

In late 2013, the EU revised the designation of phosphonates from both a fertilizer and a pesticide to that of only a pesticide. A new MRL for phosphonate-containing products was not established and these products were included in the definitions of the existing MRL for the pesticide fosetyl-al. Fosetyl-al is not authorized for use on fruit bearing trees in the US, but phosphonate-containing products are sold both as registered fungicides and as foliar fertilizers. The US EPA exempts potassium phosphite fungicides from requiring a MRL and in the US, fertilizers do not require MRLs. Hence, no MRL or studies to support a MRL for phosphonate-containing[1] fertilizer products existed in the US.

This EU change in designation was treated as an enforcement action rather than a change of legislation and, hence, no formal notification was made to the WTO to inform trading partners of this change. There was no opportunity afforded to comment on potential economic impacts, a limited period for compliance or for submission of data. Due to its common use as a foliar fertilizer, US almond, walnut, and pistachio growers were suddenly in excess of the EU MRL for potassium phosphite due to its inclusion in the new 2013 definition.

A temporary MRL of 75 ppm for tree nuts and other affected crops was established by the EU, effective until the end of 2015, which the EU judged to be a sufficient period of adjustment. The temporary MRL was applicable to almonds, cashew nuts, hazelnuts, macadamia nuts, pistachios, walnuts; apricots, cherries, peaches, plums and other stone fruit; blackberries, dewberries, raspberries and other cane fruit; blueberries, currants, gooseberries, figs, kumquats, persimmon, passion fruit, papaya, pomegranates, garlic, beans, peas, and asparagus.

US nut growers began residue trials under the IR-4 minor use program[2] but had to wait for a growing season in order to time the applications properly and await the analyses. Conducting the residue trials and moving the dossier through the EU regulatory approval system requires several years. The US tree nut industry submitted an information package to the EU detailing its efforts to address the new MRL requirements, including a timeline for an import MRL application. As a result of this package, the US tree nut industry was granted an extension of the temporary MRL of 75 mg/kg until the end of February 2019. Other industries that were unable to provide additional data or demonstrate similar levels of progress on field trials for an import MRL did not receive an extension. The EU returned the MRL to 2 ppm and expected farmers to limit the use of phosphonate containing products by the end of 2015.

Scott (2016) reports that on January 1, 2016, the EU MRL reverted to the detection level set at 2 mg/kg. The US's ability to export berries (and very likely, other commodities that use fosetyl or other phosphonate crop inputs) has been affected by the return to this MRL level.

Nearly US$100 million of US fresh and dried fruit and berry exports (including US$70 million of dried prunes alone) are subject to the MRL (of 2 ppm) as of January 1, 2016. US exports are often only a small part of the EU's overall imports of a given commodity; other countries may face even greater challenges because of the EU's reversion to the default MRL for these commodities. The value of these same products exported by other countries to the EU, which became subject to the reverted to MRL in January 2016, was approximately US$761 million in 2014. European consumers and EU agri-food supply chains will face shortages and/or suffer price increases as international suppliers adapt to the reverted-to MRL.

Had the extension for tree nuts not been obtained, the immediate impact that this change in designation could potentially have been of significant importance for the US industry. Based on 2009 data, for example, California alone exported 70% of its almond production, (30% of which are destined to the EU), 57% of its pistachios, and 45% of its walnuts.

The EU has, subsequently, agreed to establish a forum to discuss harmonization of MRLs with the US in the Transatlantic Trade and Investment Partnership negotiations.

The sources for this case study are: California Walnut Commission (2014), Vandercammen (2016), USDA (n.d.), and Scott (2016).

6.3 Cocoa from Ghana

Cocoa is Ghana's major cash crop, accounting for 25% of total foreign exchange earnings annually and supporting much of the rural economy. Between 2000 and 2009, cocoa maintained a 32% share of the value of Ghana's total exports. In 2010, exports of cocoa beans alone accounted for US$1.66 billion. In 2013, the total value of Ghana's cocoa bean exports was US$1.38 billion, with nearly US$960 million destined for the EU, just under US$65 million exported to Japan, and US$123 million going to the US.

The importance of Ghana's cocoa exports is such that the Ghana Cocoa Board adopted the Japanese Ministry of Health, Labour and Welfare's (MHLW) methodology in pesticide residue analysis for organochlorines and organophosphates for cocoa beans. Ghana complies with the Japanese regulation, which analyzes whole cocoa beans, and consequently, is more likely to result in residue violations (the EU and US de-shell the beans prior to analysis); meeting Japan's MRL for cocoa beans qualifies as meeting the US and EU MRLs as Japan's is the most stringent. Despite these proactive efforts to ensure minimal disruption to their cocoa exports, Ghana has encountered several new MRL-related problems.

Ghana has been establishing MRLs for cocoa beans destined for export based on Japan's pesticides of interest. These include chlorpyrifos, fenvalerate, pirimiphos-methyl, endosulfan, promecarb, and fenitrothion. However, these particular pesticides are no longer authorized for use on cocoa in the EU under the 2008 EU regulation on pesticide residues on imported foodstuffs. This has now caused a difficulty for Ghana's cocoa exporters who must now prioritize which export market that they wish to supply.

Ghana produced 750,000 metric tons of cocoa beans between 2008 and 2010. Of these, 40% were tested for pesticide residue and only those meeting the Japanese MRL standard were exported. These beans are re-tested upon arrival in Japan. Despite Ghana only shipping beans that were pre-tested to meet Japan's MRL, 113 residue violations occurred in testing post-arrival in Japan between 2008 and 2010.

The source of these violations was discovered to be discrepancies in the pesticide residue analysis methodology between the actual Japanese MHLW methodology utilized in Japan and the Japanese MHLW methodology adopted by the Ghana Cocoa Board. A study found that in Japan, the methodology and instrumentation used for pesticide residue analysis underwent a continuous upgrade. The upgraded methodology and process conducted upon Ghanaian cocoa beans upon arrival in Japan have relatively lower detection limits for pesticide residues.

Hence, the concern for Ghana is that the made-in-Japan methods used in Ghana were not sufficient to detect residues that the current Japanese methods can, despite their diligence in complying with the Japanese standards.

Such requirements for continuous upgrades in instrumentation and chemicals pose a significant barrier for exporting countries with limited resources. They face an endless loop of ever-increasing standards, specific to one market and conflicting with other markets, which a developing country such as Ghana can ill afford to keep pace with. Jonfia-Essien (2012) concludes with the admonishment that:

> to alleviate the bottomless(ness) in MRL of pesticide residues in cocoa beans, with all the pitfalls, all methodologies must be harmonized. The methodologies developed in Japan, EU, and USA must be harmonized into one standardized model. Such a method should be used for all confirmatory pesticide residue analysis on cocoa beans (p. 142).

The sources for this case study are: Essegbey and Ofori-Gyamfi (2012) and Jonfia-Essien (2012).

6.4 Cranberries and the EU

In 2015, total cranberry imports into the EU were valued at just under €290 million, with Canada and the US being the main suppliers. In the period 2013–2015, the EU imported between US$20–23 million of cranberries from Canadian producers. Canada is the world's second largest producer of cranberries; Canadian total global cranberry exports in 2015 totalled nearly US$300 million.

Table 6.1 Chlorothalonil (Bravo) MRLs (ppm) for application to cranberries (2015)

US	*Canada*	*EU*	*Aus/NZ*	*Codex*	*Taiwan*
5	2	0.01	10	5	1

Source Wilson (2015)

In April 2014, the EU notified the WTO of a change in its legislation to reduce the MRL for chlorothalonil (Bravo), a fungicide that prevents fruit rot in cranberries from 2 to 0.01 ppm for exporting countries, with enforcement to begin May 2015. At the time, no comments were registered with the WTO against this EU measure. Once made aware of the change, the reduction in the EU MRL threatened US$235 to US$240 million in cranberry exports to the EU. While Canadian producers already comply with a more stringent domestic MRL for chlorothalonil than their US counterparts, the Canadian MRL is still higher than the EU's and Canadian exporters would also not be able to meet the new lower EU MRL. Table 6.1 provides a comparison of the MRL for chlorothalonil on cranberries across several jurisdictions in 2015.

While the US and Canadian cranberry industries worked with US and EU officials prior to the new MRL being implemented, US cranberry growers who planned to export to the EU refrained from using chlorothalonil during the 2015 season. Growers faced adjustment costs in switching to a new fungicide or finding different markets for their cranberries. Cranberry crops faced increased disease pressure and issues with resistance to application of alternative products.

In February 2016, after reviewing submitted material, the EU revised the import tolerance for chlorothalonil (Bravo) on cranberries from the 0.01 ppm default MRL to 5 ppm, allowing the product to again be used by the industries in the exporting countries.

The sources for this case study are: AAFC (n.d.), Agri-view (2015), EFSA (2015b), USDA (2016) and Wilson (2015).

6.5 Cherry Imports into Taiwan

Taiwan has been establishing its own regulatory system for MRLs since 1999. The current system has a limited number of MRLs; despite new ones being added, many are still missing as a backlog of MRL applications waits for approval. Taiwan does not use crop groupings[3] nor does it maintain default tolerances. Taiwan is not a member of Codex and will not defer to Codex MRLs on the basis of Taiwan having a unique diet and its political situation vis-à-vis membership in international organizations. Taiwan maintains a rigorous sampling regime for pesticide residues which places many commodity groups at risk of violations. Repeated violations will result in restrictive sanctions through the use of increased inspection rates.

In February 2016, the Taiwan Food and Drug Administration (TFDA) imposed an increased inspection rate of 40% on all imports of US cherries for 2016 in response to 16 residue violations of its MRLs in 2015. Taiwan had proposed MRLs in place for 15 of the 16 violations, and permanent acceptable cherry MRLs were in place within weeks of the findings, but Taiwan refused to consider the imports acceptable. Importers with multiple violations face 100% inspection. The following terms were imposed:

1. For those importers with no violation record in 2015, any five consecutive satisfactory shipments, each in any quantity, would lead the importer to the lower inspection tier (2–10%).
2. For those with one violation record, the importers need to have five shipments reaching three times the violative quantity.
3. For those with two-plus violations (at 100% inspection rate), five satisfactory shipments reaching three times of the quantity of the last violation would lead to the lower tier (20–50%), and then, five satisfactory shipments, each at any quantity, would bring the inspection rate for U.S. cherries imported by that specific importer down to the normal rate (2–10%).
4. TFDA promised to keep the testing time within 3 working days.
5. Shipments selected for testing may be moved to the importer's warehouse with an affidavit that they will not be moved or sold before a satisfactory test result is obtained.

The value of US exports of fresh cherries to global markets ranged between US$4.5 to US$5 billion between 2011 and 2015; Taiwan accounted for US$180 to US$213 million during this period. In 2014, Taiwan imported US$213 million of fresh cherries from the US, but this declined to US$195 million in 2015.

While US cherries provide a specific example, global agricultural exports to Taiwan from many countries are affected. Taiwan's list of needed MRLs along with aggressive MRL testing has led to numerous MRL violations. Other countries in the region often look at Taiwan's violations and use it to guide their testing as well. As illustrated by the cherry example, Taiwan will penalize an entire industry for violations of individual shippers.

The sources for this case study are: Lantz (2010), Northwest Horticultural Council (2016), and USDA (2012; n.d.).

6.6 Thiram and Avocados

Thiram (tetramethylthiuram disulfide) is a type of dithiocarbamate fungicide used as a seed protectant and foliar treatment on fruit, vegetable, ornamental, and turf crops. It is also used to protect harvested crops from deterioration due to fungal infection during storage and transport. Thiram does not currently have a MRL established at Codex for use on any food product. Thiram is used on avocado crops to prevent fungal infections by many avocado producing countries. Table 6.2 presents the current MRLs for Thiram on avocados as of April 10, 2016 and indicates a large variance in the MRLs.

Thiram can be expressed as thiram, or as dithiocarbamate expressed as carbon disulfide (CS_2). When petitioned to establish an import tolerance for thiram in the EU, the MRL for both dithiocarbonate and thiram had to be evaluated. The EFSA proposed 15 mg/kg for thiram and 7 mg/kg for dithiocarbamate (CS_2), based on the authorized use for avocados in Australia. Yet, Australia and New Zealand both base their MRL on CS_2, not thiram. Specific thiram MRLs are not in place in Australia and New Zealand; nor in Mexico, the world's largest avocado producer where the use of thiram on avocados is not yet authorized.

Hence, Table 6.2 serves to illustrate the complexities of MRL establishment as Australia and New Zealand have a MRL for the use of the

Table 6.2 Thiram MRLs (mg/kg) for application to avocados (April 2016)

Australia	*EU*	*Hong Kong*	*New Zealand*	*Switzerland*	*Taiwan*	*US*	*Codex*
7	7	2	7	0.05/0.10	2.5	15	n/a

Source New Zealand Food Safety Authority—Pesticide MRL database

CS_2 form of thiram on avocadoes, which the EU subsequently based the establishment of its import tolerance upon. Yet, Australia and New Zealand do not have MRLs established for the other form, thiram. The US initially established its MRL for thiram applied to avocados at 8 ppm in December 2014, which is more reflective of other countries' MRL, (one could speculate likely based upon the CS_2 form), but revised the MRL upwards to 15 based on available data and the OECD Tolerance Calculation Procedures, effective June 2015. Again, one could speculate that the revision upwards is based on the thiram form of the pesticide. This case illustrates the impact differences in data parameters can make when trying to establish harmonized MRLs.

In 2013, 4.5 million avocadoes were harvested globally, of which 25% are internationally traded. The US is the world's largest importer, followed by The Netherlands, which acts as a transit country for avocado shipments across the EU. The top exporting countries for avocadoes in 2013 were Mexico, Chile, and Peru, with Mexico accounting for nearly half of all internationally traded avocados, valued at just under US$1 billion in 2012. Of this, 75% are destined for the US. Japan is the second and Canada the third largest market. Mexico exports nearly 30% of its total avocado production. Even though delays and/or rejections have not yet occurred in global avocado trade due to thiram-related issues, this MRL divergence poses a challenge to the world's avocado producers as they must isolate or segregate their production and shipping practices according to the MRL of the destination market(s) of their choice. Not all countries will have established a thiram MRL on avocados. The Codex MRL is missing, which creates further complexity as exporters face the dual uncertainty of how importing countries will respond to the missing MRL, both nationally and without the Codex MRL to fall back upon.

The sources for this case study are: Fresh Plaza (2013), Government of the United States (2015), New Zealand Food Safety Authority (2016), SGS (2015), and EFSA (2015a).

6.7 Snapshot of Additional MRL Cases

The following tables provide a snapshot of additional MRL cases for select importing markets. This is included to demonstrate how often violations occur, the diverse range of agricultural products affected, and the distribution of exporting countries. The latter spans both developed and developing countries. The tables do not necessarily differentiate between violations that occurred due to misaligned MRLs, a missing MRL or to a true MRL violation where GAP standards were not attained. This is a result of individual countries' reporting systems which may or may not provide such detailed data. Where possible, the cause of the violation was investigated to determine whether misaligned or missing MRLs could be the reason. Table 6.3 presents information for Australia.

Table 6.3 Australia chemical test results and random monitoring of food for compliance Dec 2015

Date	*Source*	*Product*	*Residue (mg/kg)*	*Australia MRL (mg/kg)*
10/12/2015	China	Dried red dates	Carbendazim (0.2),	MRL missing for dried red dates
			Difenoconazole (0.4),	MRL missing for dried red dates
			Tebuconazole (0.5)	MRL missing for dried red dates
16/12/2015	China	Red dates	Carbendazim (0.11),	MRL missing for red dates
			Cyhalothrin (0.06)	MRL missing for red dates; MRL for stone fruit 0.5
16/12/2015	Vietnam	Frozen chilli pepper	Carbendazim (0.14),	For dry chili peppers—20; For peppers 0.1
			Difenoconazole (0.078),	MRL missing for any sort of chili
			Propiconazole (0.095)	MRL missing for any sort of chili
17/12/2015	China	Fresh sugar snap peas	Thiamethoxam (0.1)	MRL missing for peas or legumes; for beans 0.2 and fruiting vegetables 0.5, both of which are temporary as of June 1, 2016.

Source Government of Australia (2015; n.d.)

Australia establishes its own MRLs for both domestically produced and imported food. If a particular chemical-commodity combination does not have a unique MRL, zero tolerance (no detectable limit) applies. In Australia, carbendazim is authorized for use on a variety of fruit, vegetables, meats, nuts, spices, pulses, and rice, but it is not authorized for use on red dates in any form; the two red date violations listed could likely be attributed to the missing MRL. Carbendazim is authorized for use in Australia on dry chilli peppers (20 mg per kg) and peppers (0.1 mg per kg, the limit of determination). The Vietnamese violation for frozen chilli peppers listed above could be due to the fact that the product is frozen or because Australia's MRL for peppers of 0.1 mg/kg was exceeded—which could either be due to a misaligned MRL or because GAP was not practiced. For comparison, carbendazim is not authorized for use on any product in the EU, while the Codex matches Australia's 20 mg/kg for dried chillis. The Codex MRL for chilli peppers of 2 mg/kg is greater than Australia's.[4]

Cyhalothrin is authorized for multiple uses on vegetables, grains, fruit, seeds, meats, pulses, and teas in Australia as well as for use on stone fruit at 0.5 mg/kg. The violation for red dates from China could be due to a missing MRL (if red dates are not included as stone fruit). If red dates are considered a stone fruit, the Australian MRL for cyhalothrin on stone fruit was not exceeded; hence, the violation could be solely based on the cabendazim MRL. For comparison, Codex has MRLs for 'berries and other small fruits' at 0.2 mg/kg and 'spices fruits and berries' at 0.3 mg/kg.

Through such simple examples, the uncertainty of how MRLs affect trade can be illustrated. When shipping to Australia, red dates and stone fruits do not have a MRL for carbendazim. Is the existing carbendazim MRL for a dried chilli pepper applicable to frozen chillies? If the dried chilli MRL does not apply, could the fresh chilli pepper MRL apply? If not, then zero tolerance likely applies. In these examples, limitations in data make it difficult to determine whether the MRL violation occurred due to a misaligned MRL or whether GAP standards were actually not followed.

Table 6.4 presents cases for Japan, Table 6.5 for Taiwan and Table 6.6 for the European Union.

Table 6.6 shows that the US experienced several violations for carbaryl used on minneolas (a type of tangelo) in March 2016. The violations

Table 6.4 Japan pesticide residue violations on imported food February 2016

Date	*Source*	*Product*	*Pesticide, residue level and Japan MRL status*
Feb 2016	Ghana	Fresh cacao beans	fenvalerate (0.51 ppm) Missing fenvalerate MRL for cacao or cacao beans. Dried beans have a MRL of 0.5
	Vietnam	Frozen food preheated	furzolidone 0.001 ppm This chemical was not found in Japan's database
	India	Whole cumin	profenofos 0.33 ppm Missing MRL for cumin Japan's MRL is 0.05 for 'other spices'
	Thailand	Fresh durian	metalaxyl and mefenoxam Missing MRL for durian Durian is missing as a listed agricultural product
	Ecuador	Fresh cacao beans	24D 0.02 detection Missing MRL for 24D on cacao
	Venezuela	Fresh cacao beans	24D 0.02 detection Missing MRL for 24D on cacao
	Italy	Powdered chicken egg white	furazolidone 0.027 ppm This chemical was not found in Japan's database

Source Government of Japan (n.d.a) and (n.d.b)

could have occurred for two reasons. As minneolas do not have a specific MRL for carbaryl in the EU, the MRL is missing, resulting in the listed violation as an 'unauthorized substance'. The second is, given that minneolas are not specifically listed but are a citrus fruit, if they were categorized as citrus upon inspection, the carbaryl residue exceeded the EU MRL for citrus. Table 6.7 compares carbaryl MRLs across various countries, as applied to citrus products, as a group or as individual fruit types. It shows that the US violations listed in Table 6.6 were the result of a missing MRL in the specific case of minneolas and misaligned MRLs in the general case of citrus as the carbaryl residue levels found on the minneolas exported to the EU, which only lists carbaryl for use on grapefruit and oranges, were far below the US MRL for carbaryl on citrus.

The sources for this case study are Government of Australia (2015; n.d.), EU (2016; n.d.), Codex (n.d.), Government of Japan (n.d.a.; n.d.b.), Government of Taiwan (n.d.), and Ritenour (2016).

Table 6.5 Taiwan March 1–23, 2016—imported food violations

Date	*Source*	*Product*	*Residue*
March 23	India	Sesame seed	Pesticide residue content
	India	Black sesame seed	Pesticide residue content
	USA	Fresh grapefruit	Pesticide residue content
	France	Assorted red fruit	Pesticide residue content
March 22	USA	Ground cumin	Pesticide residue content
March 15	Israel	Fresh rosemary	Pesticide residue content
	Thailand	Leafy vegetables	Pesticide residue content
	Serbia	Frozen raspberries	Pesticide residue content
	Japan	Kiwifruit	Pesticide residue content
	Vietnam	Green tea	Pesticide residue content
	Japan	Fresh radish	Pesticide residue content
	Kenya	Fresh chervil	Pesticide residue content
March 8	Japan	Fresh perilla green leaf basil	Pesticide residue content
	Chile	Fresh blueberries	Pesticide residue content
	India	Brown sesame seed	Pesticide residue content
	India	Brown sesame seed	Pesticide residue content
	Kenya	Fresh rosemary	Pesticide residue content
	Italy	Fresh arugula	Pesticide residue content
	Netherlands	Fresh leeks	Pesticide residue content
	Thailand	Fresh asparagus	Pesticide residue content
March 04	Canada	Organic unsweetened cranberry apple juice	Pesticide residue content
	Chile	Fresh blueberries	Pesticide residue content
	Japan	Half green tea powder	Pesticide residue content
	Thailand	Holy basil	Pesticide residue content
	Thailand	Kaffir lime leaves	Pesticide residue content
March 1	Pakistan	Edible dried rose petals	Pesticide residue content
	India	Red hemp (brown sesame seeds)	Pesticide residue content
	Japan	White plums	Pesticide residue content
	China	Dried hericium	Pesticide residue content
	Japan	Satsuma mandarin	Pesticide residue content
	Japan	Komatsu vegetables	Pesticide residue content
	Japan	Fresh garland chrysanthemum	Pesticide residue content
	China	Cantonese style pepper (food use input)	Pesticide residue content
	India	Green cardamom	Pesticide residue content
	USA	Fresh salad package baby arugula and greens	Pesticide residue content
	Japan	Fresh perilla basil leaf	Pesticide residue content
	Japan	Fresh citron	Pesticide residue content
	Italy	Fresh fennel	Pesticide residue content
	Chile	Fresh blueberries	Pesticide residue content

(continued)

Table 6.5 (continued)

Date	*Source*	*Product*	*Residue*
	Vietnam	Fresh sweet peas (no shell)	Pesticide residue content
	Japan	Fresh mikan citrus	Pesticide residue content
	Kenya	Fresh basil	Pesticide residue content
	Japan	Seasoning powder	Pesticide residue content
	China	Fresh chestnuts	Pesticide residue content
	India	Brown sesame seeds	Pesticide residue content
	Japan	Fresh sour orange	Pesticide residue content
	USA	Fresh spinach salad pack	Pesticide residue content
	USA	Fresh salad pack	Pesticide residue content
	USA	Fresh kale salad pack	Pesticide residue content
	Japan	Fresh basil leaves	Pesticide residue content

Source Government of Taiwan (n.d.)

6.8 MRL Violations on Generic Pesticide Products

When a pesticide's patent expires, generic versions of it can be produced and sold by companies other than the originating patent holder. Although generic products have the same active ingredient(s), they are usually sold at discounted pricing from that of the branded product; hence, generic products provide a cost-effective alternative pesticide choice for producers of many crops around the world. In 2015, generic pesticides (or post-patent materials) accounted for 45% of the total market value of global agri-chemicals, which was estimated to be just over US$60 billion at the manufacturer level in 2013 (Xie 2015).

The research and development costs of generic pesticide manufacturers are significantly less than the original patent holders, because they did not have to invent the product. They still must test their generic version and seek government review, approval/registration for it. A generic manufacturer essentially manufactures the active ingredients/substances of pesticides, the patents of which have expired, and does not hold the original patents. They also do not necessarily have access to the original data submission package used to register the patented pesticide. Generic pesticides, by their nature, utilize older active ingredients.

Generic pesticides are especially challenged by MRL requirements. They are older products, with many firms putting them on the market. Their residue and data packages are older and out of date, if they exist at all, and are for the purposes of applying for a MRL in 2017, largely

Table 6.6 EU imported food notifications* March 1–25, 2016

Date	*Source*	*Product*	*Issue*	*Action*
Mar 25	Spain	Lemons	Prothioconazole 0.41 mg/kg-ppm	Alert EU MRL 0.02
	Turkey	Fresh sweet peppers	Lambda-cyhalothrin (0.26 mg/kg-ppm)	Border rejection EU MRL is 0.1
Mar 23	Turkey	Grapefruit	Imazalil (11 mg/kg-ppm)	Info for attn. EU MRL is 0.1
Mar 21	USA	Minneolas	Unauthorized substance-carbaryl (0.54 mg/kg-ppm)	Info for attn. EU MRL is 0.01 for grapefruit and mandarins
Mar 16	USA	Minneolas	Unauthorized substance-carbaryl (0.2, 0.57 mg/kg-ppm)	Alert
Mar 14	Turkey	Fresh cucumbers	Chlorpyrifos (0.138 mg/kg-ppm)	Border rejection EU MRL is 0.05
	USA	Minneolas	Unauthorized substance-carbaryl (0.57 mg/kg-ppm)	Alert EU MRL is 0.01 for grapefruit and mandarins
	Turkey	Sweet peppers	Methomyl (0.078 mg/kg-ppm)	Border rejection EU MRL is 0.02
Mar 10	Vietnam	Frozen catfish	Prohibited substance-nitrofuran	Border rejection
	India	Basmatic rice	Unauthorized substances acephate (0.03) and carbendazim (0.08)	Border rejection Acephate on rice 0.01 Carbendazim on rice 0.01
	India	Fermented black tea	Unauthorized substance propargite	border rejection EU MRL 0.05
Mar 4	Poland	Chilled green peppers	Ethephon (5 mg/kg – ppm)	Info for attn. EU MRL is 0.05
Mar 3	China	Green tea	Unauthorized substance – anthraquinone (0.056 mg/kg-ppm)	border rejection EU MRL is 0.02
Mar 2	Turkey	Sweet peppers	Fosthiazate (0.095 mg/kg-ppm)	border rejection EU MRL is 0.02

Source EU (n.d.)

*Does not include notifications for non-pesticide related issues such as but not limited to adulteration, bacterial or fungal contamination, colorants, temperature control, undeclared ingredients, mercury, and sulphites.

Table 6.7 Global MRL comparisons for carbaryl used on citrus, ppm

Chemical	*Common trade name*	*US citrus*	*Canada citrus*	*Codex citrus*	*EU (grapefruit & orange only)*	*Japan (grapefruit only)*	*Taiwan (grapefruit & orange only)*	*South. Korea (grapefruit & orange only)*
Carbaryl	Sevin	10	10	15	0.01	7	2	0.5

Source Ritenour (2016)

useless. The original patent holder would have the data package but has no economic incentive to update or revise the data. It may have sold the data. Furthermore, the company itself may no longer exist. Current individual manufacturers or sellers of the generic will generally not undertake the research needed to generate new data packages as the costs of doing so would not be justified by anticipated returns. Some generic manufacturers will further channel research resources into particularly valuable or important active ingredients for use in new crops and applications, and will, therefore, generate the data packages needed for updated registrations and MRLs. Generic pesticides may not have Codex MRLs; few if any will have individual country MRLs.

Hence, the asynchronicity of global MRLs, and the trend of moving away from Codex MRLs towards individual country MRLs, threatens the use and viability of many generic pesticides. As many generics will not have individual country MRLs, crops using the generic pesticide will not be accepted as imports in those countries, resulting in a loss of export markets for those producers using that generic product. They can choose to either discontinue their use of the generic pesticide, thereby incurring higher costs of production, but retain access to that particular country's market or continue to use the generic pesticide but find new markets for their crops. As more countries are moving to their own national MRLs, there will be fewer and fewer markets to choose from. The loss of ability to use generic pesticides will increase the costs of production for producers of many crops around the world. Many of these are in developing countries and, hence, inhibit market access and ability to trade for developing countries, with negative implications for food security.

The case study of avocadoes and thiram given above is one highlighting the generic issue as thiram has been registered for use in the US since

1948. In its mandated pesticide reregistration process for thiram, conducted in 1991 and 1995, the EPA noted missing or inadequate studies in thiram's scientific database. It required additional generic studies for thiram from several manufacturers to confirm its regulatory assessments and conclusions (EPA 2004). The lack of MRLs in other countries and at Codex for thiram's use on avocadoes can be explained by its generic status—there is no incentive for an individual company to undertake the cost of updating data and studies in order to submit a MRL registration for thiram's use on avocadoes in one, never mind multiple countries.

Another example of the problems MRLs pose for generics is provided by South Korea's reform of its MRL system. Its system prior to 2017 was based upon a *decision tree* where a national MRL, if it existed, applied. If a national MRL did not exist, a commodity-specific Codex MRL would be used. If no Codex MRL existed, a South Korean MRL established on a similar product would apply. This system had enabled agri-food shipments to flow relatively unimpeded by MRL violations, even when a national South Korean MRL for a specific chemical/commodity combination was missing (Lantz 2016).

The revised South Korean MRL system became effective January 1, 2017 and will eventually eliminate the *decision tree* system. Only South Korean MRLs will be accepted and if none exists, a default tolerance of 0.01 ppm will apply. During the transition to this new system, South Korea has requested data packages from pesticide registrants to be submitted for review and to establish MRLs at a cost of US$5000 per MRL (Lantz 2016). The first part of the positive list transition occurred as scheduled on January 1, 2017 and is applicable to nuts, oilseeds and tropical fruit. Other commodities will undergo the MRL revision by December 2018 (2016).

Generic or post-patent products are sold by numerous registrants who may or may not have the data needed to seek a MRL in South Korea. They may or may not have the technical capabilities, a physical location in South Korea, or even the funds to conduct comprehensive reviews. With the large number of registrants selling the post-patent compound, who is responsible for the making submission and who will bear the costs of doing so? Failure to make a submission will negate the South Korean national MRL, if one exists, and the default tolerance of 0.01 ppm will apply. It is estimated that up to one-third of necessary MRLs are faced with this dilemma, with no strategy to comply with a South Korean MRL (Lantz 2016).

Depending on the commodity, it is estimated that US producers alone will require new South Korean MRLs for between 50 and 90% of the crop protection products which they use. An estimated 40–80 new MRLs are needed in South Korea for each commodity. Many of these products enjoyed the import compliance of the existing system's deferral path, which will be lost under the new system unless a new South Korean MRL is established. Hence, crops and food products that were previously compliant with the South Korean system will no longer have an applicable MRL. If tested, residues could be found that result in shipments being rejected as well as potentially, increased testing in that commodity's imports, with further rejections and ultimately, reports of *unsafe* food entering South Korea (Lantz 2016).

These case studies illustrate the frequency and complexity of MRL violations and issues across a broad range of agri-food products, from a diverse set of global exporters. As more countries transition towards their own positive list of regulated MRLs, they must undertake their own research and due diligence to establish MRLs and default tolerances, a time-consuming process. This creates a backlog and a discrepancy between pesticides allowed for use in exporting countries and those that do not yet have their MRL on the positive list. This combination of a positive list and missing MRLs has resulted in increased MRL violations and is a growing and costly problem impeding the global trade of agri-food products. Countries also update and revise their MRL limits, usually with increased stringency or reduced applications, which also causes discrepancies between what used to be permitted and the new standard. This was illustrated by the several of the case studies. Exporting producers require sufficient time, in a timeline suitable for their crop, to transition to the new standard; often, however, new MRLs take effect too rapidly for exporters to adjust as illustrated in the Philippine banana and US tree nut cases.

Developing and less developed countries are particularly affected by MRL misalignment as the case study discussing cocoa exports from Ghana illustrates. In trying to expand their economies and provide greater wealth to their citizens through agricultural exports, these countries are affected by many factors which can impede that growth, but the proliferation of MRL standards and the increasing sophistication of detection methodologies around the world can be a significant problem. Limited government resources combined with under-developed technical capacity and regulatory systems make it difficult for these countries to meet the import standards of their trading partner countries. As discussed previously, the establishment of international standards at Codex and their adoption by all countries would be of significant benefit to developing countries as it would provide a means to cope with changing regulatory requirements and to enhance their food security.

However, that is not the reality and it leaves these countries with not only difficult, but also sometimes impossible choices. The greatest growth and wealth generation opportunities are generally in exporting to developed countries with sufficient income and wealth to pay a higher price for the agricultural products they demand. Developed countries generally have very sophisticated regulatory systems, highly trained technical labor, and state-of-the-art science, and therefore, have the ability to continuously develop new standards to replace old standards and rapidly establish new detection levels.

The growing tendency of countries to move away from international standards and establish their own national ones, therefore, makes it extremely difficult for developing countries to stay current with all the changes. In the case of Ghana, it chose to adopt the Japanese standard, which worked until standards changed in the EU and the US, and detection levels were lowered in Japan.

From a policy perspective, this regulatory misalignment and its impact on food security goals is a challenge for everyone. The developed countries, in general, through the work of international organizations and through providing resources, have commitments to work towards increasing the food security of developing countries and bring more people in these countries out of poverty. For agriculture, this is one of the primary goals of the FAO and its regional organizations. Therefore, when developed countries change import requirements to meet their own domestic circumstances, they can unintentionally and adversely affect agricultural development in other countries.

There may be a number means by which to address this gap; however, one of the most logical could be greater convergence at the FAO between its development and food security goals and the development of Codex standards. This book outlines the factors limiting progress on international standards at Codex and proposes some steps forward. While the focus of improving Codex has been on food safety and trade, there is an opportunity for the FAO and Codex to do more work on how international standards can contribute to food security.

Described as *death by a 1000 cuts*, MRL violations occur daily in virtually all markets as illustrated above. Each violation results in a hold on the shipment, while the necessary remedy is sought. Such holds on imports incur demurrage and storage costs. If rejected, shipments must be rerouted or destroyed. Rerouting a container costs an average of

US$20,000. Destroying a shipment also imposes significant costs on the exporting firm, and can imperil its viability. As costs associated with MRL violations increase, importers and consumers will face price increases as MRL-related costs are passed onwards and/or shortages of the product occur. Even with registrants implementing global MRL strategies, MRL coverage will still be incomplete due to the number of countries moving to national MRL lists and/or countries providing insufficient guidance to registrants trying to obtain a MRL. It takes time for a country to establish MRLs and backlogs develop as the Taiwan case study illustrates.

It is important to note that the focus and goal of this study are to address misaligned global MRL regulatory policies for agricultural commodities moving across borders that result in MRL violations. Misaligned MRL regulations and/or missing MRLs result in crops that are produced properly and legally in their domestic market being rejected at destination markets. However, MRL violations and rejections can also occur as a result of true failures in residue management by producers and exporters. These types of violations indicate that the MRL system is, in fact, operating as intended, as compliance with MRLs indicates that GAP were used. If a true violation occurs, it indicates that the resulting product is not GAP compliant. These types of violations are not the focus of this book.

While academic studies of MRL (and related) issues are examples of the trade difficulties caused by asynchronous regulations, providing considerable insights into the extent of the problem and its implications for trade and food security, they cannot provide a look into the thinking of those actually charged with developing and administering MRLs. The next section provides a step toward filling that gap by reporting the results of structured interviews with members of the regulatory bureaucracy from a variety of countries.

Notes

1. Phosphonate is a term that describes any compound containing a carbon to phosphorous bond. Phosphonate products include potassium phosphite, phosphorous acid and fosetyl-Al among others, which can act as either fungicides or fertilizers (Center for Turfgrass Science, n.d.).
2. IR4 is the Inter-Regional Research Project Number 4, also known as the Minor Crop Pest Management program which assists in the collection of residue and efficacy data in support of the registration or reregistration of minor use pesticides and the determination of MRLs for specialty crops. It is a jointly managed by the USDA and EPA.

3. Crop groupings are discussed in greater detail in Chapter 8. Crops sharing agronomic similarities can be grouped together and the data used in establishing a MRL for one crop can be considered for the entire group.
4. These comparisons were made using Australia New Zealand Food Standards Code Schedule 20; the EU Pesticides Database; and the Codex Pesticide Database.

References

AAFC. (n.d.). 'Canadian Cranberries', Agriculture and Agrifood Canada, Ottawa. http://www.agr.gc.ca/eng/industry-markets-and-trade/exporting-and-buying-from-canada/buy-from-canada/canadian-cranberries/?id=1426172190727. Accessed 6 Mar 2016.

Agri-view. 2015. Cranberry Exports Impacted by EU Decision, Jan 29. http://www.agriview.com/news/crop/cranberry-exports-impacted-by-eu-decision/article_2c5914a6-50ca-5c13-a764-7530f42efb14.html. Accessed 6 Mar 2016.

Arcalas, J. 2016. S. Korea retains rules on PHL banana exports. *Business Mirror*, December 4. http://www.businessmirror.com.ph/s-korea-retains-rules-on-phl-banana-exports/. Accessed 8 Feb 2017.

AsiaFruit. 2013. Bananas Rank Top in Korea, July 10. http://www.fruitnet.com/asiafruit/article/158854/bananas-rank-top-in-korea. Accessed 3 Mar 2016.

Bryant Christie. 2016. Implementation of Korea PLS Phase One—January 1, 2017, December 16. http://www.bryantchristie.com/blogs/implementation-korea-pls-phase-one-%E2%80%93-january-1-2017. Accessed 8 Feb 2017.

Business Mirror. 2015. Korean Rules on Pesticide Residue Threaten Banana Exports, Feb 22. http://www.businessmirror.com.ph/korean-rules-on-pesticide-residue-threaten-banana-exports/. Accessed 3 Mar 2016.

California Walnut Commission. 2014. EU Changes Definition of Phosphite Residues, Dec 30. http://www.walnuts.org/walnut-industry/announcements/eu-changes-definition-of-phosphite-residues/. Accessed 4 Mar 2016.

Center for Turfgrass Science. (n.d.). 'Fact Sheet, Phosphonate Products', College of Agricultural Sciences, Pennsylvania State University: University Park. http://plantscience.psu.edu/research/centers/turf/extension/factsheets/phosphonate-products. Accessed 4 March 2016.

Codex. (n.d.). *Pesticides Residues in Food Database*, Rome. http://www.fao.org/fao-who-codexalimentarius/standards/pesticide-mrls/en/. Accessed 3 June 2016.

Diega, A.S. 2015. DA asks South Korea to Extend Start of Stiffer Pesticide Protocol on PHL Bananas. *Business Mirror*, August 6. http://www.businessmirror.com.ph/da-asks-south-korea-to-extend-start-of-stiffer-pesticide-protocol-on-phl-bananas/. Accessed 3 Mar 2016.

EFSA. 2015a. Reasoned Opinion on the Modification of the Existing MRL for Thiram in Avocados. *EFSA Journal*, January. http://www.efsa.europa.eu/en/efsajournal/pub/4003.

EFSA. 2015b. Setting of an Import Tolerance for Chlorothalonil in Cranberries, European Food Safety Authority, Brussels, July 2. http://www.efsa.europa.eu/sites/default/files/scientific_output/files/main_documents/4193.pdf.

EPA. 2004. RED Facts—Thiram, Reregistration Eligibility Decision Document, Environmental Protection Agency, September. http://pmep.cce.cornell.edu/profiles/fung-nemat/tcmtb-ziram/thiram/thiram_rer_0904.pdf. Accessed 12 Feb 2017.

Essegbey, G., and E. Ofori-Gyamfi. 2012. Ghana Cocoa Industry—An Analysis from the Innovation System Perspective. *Technology and Investment*, 3: 276–286. http://www.worldcocoafoundation.org/wp-content/uploads/files_mf/essegby2012socioeconomicinnovation214kb.pdf.

EU. 2016. *EU Pesticides Database*, Brussels. http://ec.europa.eu/food/plant/pesticides/eu-pesticides-database/public/?event=activesubstance.selection&language=EN.

EU. (n.d.). *RASFF Chronological List*. https://webgate.ec.europa.eu/rasff-window/portal/?event=notificationsList&StartRow=1.

Fresh Plaza. 2013. Avocado Market Still Growing, Holland Plays a Big Role in It, June 13. http://www.freshplaza.com/article/110214/Avocado-market-still-growing.-Holland-plays-big-role-in-it. Accessed 9 Apr 2016.

Government of Australia. 2015. Failing Food Report—December 2015, Department of Agriculture and Water Resources. http://www.agriculture.gov.au/import/goods/food/inspection-compliance/failing-food-reports/december-2015#random–monitoring-of-food-for-compliance. Accessed 6 Mar 2016.

Government of Australia. (n.d.). Australia New Zealand Food Standards Code—Schedule 20—Maximum residue limits. *Federal Register of Legislation*. https://www.legislation.gov.au/Details/F2016C00545. Accessed 6 Mar 2016.

Government of Japan. (n.d.a). Recent Cases of Violation of the Food Sanitation Law that were Found on the Occasion of Import Notification. //www.mhlw.go.jp/english/topics/importedfoods/. Accessed 8 Mar 2016.

Government of Japan. (n.d.b). Positive List System for Agricultural Chemical Residues in Foods. http://www.m5.ws001.squarestart.ne.jp/foundation/search.html. Accessed 25 June 2016.

Government of Taiwan. (n.d.). 'Imported food violations', Food and Drug Administration. https://consumer.fda.gov.tw/Food/UnsafeFood.aspx?nodeID=170. Accessed 28 Mar 2016.

Government of the United States. 2015. Thiram; Pesticide Tolerance, A Rule by the EPA on 06/19/205, *US Federal Register*, Washington. https://www.federalregister.gov/articles/2015/06/19/2015-14944/thiram-pesticide-tolerance.

Jonfia-Essien, W.A. 2012. Emerging International Standards in Cocoa Trade: Recent Treatise on MRL. In *Proceedings of the 9th International Conference on Controlled Atmosphere and Fumigation in Stored Products*, eds. H.J. Banks, D.S. Jayas, C.H. Bell, R.T. Noyes, A.G. Ferizli, M. Emekci, A.A.

Isikber and K. Alagusundaram, Oct 15–19, 136–143, Antalya. http://ftic.co.il/2012AntalyaPDF/SESSION%2002%20PAPER%2003.pdf.

Lantz, M. 2010. *Pacific Rim MRL Issues.* Presentation to California Special Crops Council International MRL and SPS Issues Workshop, June 2. http://www.specialtycrops.org/MRL_pdfs/5%20MRL%20Issues%20in%20Asia.pdf.

Lantz, M. 2016. The Coming MRL Challenge in Korea. Seattle: Bryant Christie Inc., March. https://www.globalmrl.com/downloads/whitepaper_mrl_challenge_korea.pdf.

New Zealand Food Safety Authority. 2016. *Pesticide MRL Database.* Wellington: Government of New Zealand. https://pxmrl.maf.govt.nz/Residues/Pesticide/THIRAM/Crop/Avocados/Default.aspx.

Northwest Horticultural Council. 2016. Taiwan Cherries: Increased Residue Testing, March 10, Country Alerts. http://nwhort.org/taiwan-cherries-increased-residue-testing/. Accessed 7 Mar 2016.

Ritenour, M.A. 2016. *2016 Florida Citrus Pest Management Guide: Ch. 3 Fresh Fruit Pesticide Residue Limits.* University of Florida IFAS Extension, Publication #HS-1124. http://edis.ifas.ufl.edu/cg087.

Scott, J. 2016. *Foreign Agricultural Service's Role in Addressing Trade-Related MRL Issues.* USDA presentation to US Specialty Crops. http://specialtycrops.org/MRL_pdfs/2013/4%20Scott%20USDA-FAS%20%20%20%20Role%20in%20MRL%20Issues.pdf.

SGS. 2015. US EPA Revises MRL for Thiram in Avocadoes, June 30. http://www.sgs.com/en/news/2015/06/safeguards-10715-us-environmental-protection-agency-epa-revises-mrl-for-thiram-in-avocadoes Accessed 9 Apr 2016.

USDA. 2012. US Specialty Crops Trade Issues 2012. In Annual Report to Congress. Washingon: United States Department of Agriculture. http://www.fas.usda.gov/sites/default/files/2014-03/7922513_enclosure.pdf.

USDA. 2016. 'Chlorothalonil MRL on Cranberries at 5 ppm Again', US Mission to the EU, Foreign Agricultural Service, United States Department of Agriculture, Brussels. http://www.usda-eu.org/chlorothalonil-mrl-on-cranberries-at-5-ppm-again/. Accessed 6 Mar 2016.

USDA. (n.d.). Global Agriculture Trade Statistics, Foreign Agricultural Service.

Vandercammen, G. 2016. *New EU MRL for Fosetyl Takes Effect.* USDA FAS GAIN Report E16001, Washington, Jan 4. https://gain.fas.usda.gov/Recent%20GAIN%20Publications/New%20EU%20MRL%20for%20Fosetyl%20Takes%20Effect_Brussels%20USEU_EU-28_1-4-2016.pdf.

Wilson, J. 2015. *Horticultural Research and Regulatory Activities.* The Cranberry Institute presentation to the BC Cranberry Congress, Feb. http://www.bccranberries.com/pdfs/2015_cranberry_congress/4-Revised-Wilson.pdf.

Xie, C. 2015. Four Billion Dollar Worth of Pesticides to Come Off Patent in 5 Years. *Agronews,* March 12. http://news.agropages.com/News/NewsDetail—14166.htm. Accessed 3 June 2016.

CHAPTER 7

What Do Administrators Say?

Abstract The results of a survey of individual government officials charged with administering MRL policy are reported. Respondents from a variety of countries with divergent policy regimes for MRLs were interviewed. They were asked about their current practices related to MRLs and how progress might be made in achieving regulatory harmonization for MRLs. Their answers were informative regarding both regulatory divergence and how barriers to trade could be reduced.

Keywords Interviews · OECD calculator · Joint reviews · Codex MRL · Crop groupings · Data package requirements · Technical differences · Resource scarcity · Technical capacity

In-person interviews and discussions pertaining to misalignment of MRLs were conducted with regulators in a variety of countries. They were asked their government's perspective on the use of Codex MRLs, the OECD calculator,[1] joint reviews, problems experienced with exports and imports due to MRLs, and suggestions for MRL harmonization. The sample is small and the participants were not randomly selected; they were selected to represent an array of different perspectives. While the responses must, as a result, be treated with degree of caution, they still provide insights into the thinking of regulators. A copy of the questionnaire can be found in Appendix 1.

M.T. Yeung et al., *Declining International Cooperation on Pesticide Regulation*, Palgrave Studies in Agricultural Economics and Food Policy, DOI 10.1007/978-3-319-60552-4_7

Given the sensitivity of some of the issues, the interviews were conducted on the condition that confidentiality would be respected. As a result, descriptive identifiers are used in the discussion below:

1. Country A—An agri-food trade-dependent (both exports and imports) developed country;
2. Country B—A tropical fruit exporting developing country;
3. Country C—An agri-food net importing developed country;
4. Country D—An agri-food net exporting developed country.

The results of the interviews are reported in the order of the issues raised in the questionnaire.

1. Use of internationally agreed Codex MRLs

   The interview responses show that Codex MRLs are variously used but only to supplement a country's own national MRL system in some way and are not usually deferred to. Even when the country is missing a specific MRL, the Codex MRL may not be deferred to and zero tolerance becomes the default. Codex MRLs and processes can be considered when countries establish their own MRL for a product or when a specific request to align MRLs is received. It would seem that while Codex MRLs are not often deferred to directly, they and/or their data packages are given some form of consideration in the interests of facilitating some harmonization with other countries.

   a. Country A—while they participate in the Codex process, they have their own national MRLs, set by their own regulatory review process. They are unlikely to defer to Codex MRLs, but will consider other countries' data during the assessment process so as to achieve a degree of harmonization.
   b. Country B—does have national MRLs established, but they are done by crop and focused on exports. The government works with the pesticide registrant when approving a domestic registration for use on a domestic crop. They will consider Codex data packages when making their decisions and often make use of information on the US EPA website. When they allow for a registration, they will also set a national MRL. Codex will be considered when setting the MRL, but also regional MRLs, and

potential export markets' MRLs. This can result in the lowest MRL being established depending on what the standards are in the export market. Regional MRLs are considered but not necessarily formally deferred to—there is an informal exchange system depending on which crop regional partners specialize in.

c. Country C—They will consider a Codex MRL when they are establishing a national MRL, but do not defer to Codex. They must complete their own review process to establish either registrations or import tolerances, but when they are doing their reviews, they will look to see if a Codex MRL has been established. Potential future use of Codex MRLs might be considered for something that is not reviewed or approved, but resource issues prevent this. For Codex to be improved, the JMPR[2] needs more resources.
d. Country D—Codex MRLs are used only in some circumstances; they are recognized as having a role in international trade; hence, requests to align with Codex or other recognized MRLs may be accepted through a proposal process if all requested information is provided and specific criteria are met. Otherwise, this country establishes is own MRLs for food and Codex MRLs are not referenced, though JMPR evaluations can be used in a supportive role if their data package is submitted. They use their own country-specific GAP, data, and use patterns in combination with internationally accepted risk assessment methodologies to establish their national MRLs.

2. The OECD MRL calculator is used mainly by respondents that are members of the OECD. This corresponds with other sources that indicate use of the calculator is not very common in Asia and non-OECD countries. However, use of the calculator, even amongst OECD members, can be hampered by language and differences in data requirements between the calculator and their own national processes.

    a. Country A—yes, is a committed user and made contributions towards resolving data set issues;
    b. Country B—not used;
    c. Country C—not directly used due to discrepancies in field trial data. However, when import tolerances are being set, the submitted applications seek a MRL based on the OECD

calculator. They do consider foreign MRLs when establishing national MRLs. If possible they do try to harmonize MRLs with Codex or to the country making the application.

d. The OECD calculator is used as a tool to inform the decision of their national MRL—the calculator's output is considered in the context of the data set and can be used in the decision.

3. The use of crop groupings is not universal amongst respondents; however, the two countries that do not currently use crop groupings are aware of, and expressed potential willingness to consider, using crop groupings at some future point. The two countries currently using crop groupings do so to varying degrees, from one that actively uses crop groupings to the other that will use their own unique crop groupings but not Codex groupings, although they will use Codex groupings as a basis for MRL establishment.

   a. Country A—yes, crop groupings are actively used and are active in crop groupings work.
   b. Country B—though the use of crop groupings would be considered, it is far less a priority than setting the national MRL for a crop being grown for export.
   c. Country C—not traditionally but are monitoring Codex crop groupings. There is a possibility in the future that crop groupings would be considered, but would be adapted to their unique needs.
   d. Country D—uses crop groupings if the use of the chemical associated with the MRL requires a group MRL; Codex crop groupings are used only as a basis for MRL establishment. This country maintains its own crop groupings list, the contents of which describes the food commodities and the proportion of the commodity to which the MRL applies. Their list is a subgroup of commodities listed by Codex and is specific for use in their country; if the requested commodity is included in the Codex crop group but not the country's crop grouping, separate MRL requests are required. Their legislation precludes the automatic adoption of or reference to Codex groupings in their regulations.

4. Participation in joint reviews from respondents is mixed, depending on available resources, including the availability of qualified,

technical labor. Participation is often limited by language and technical differences between a country's domestic MRL regime (particularly in data package specifications) compared to OECD or Codex.

a. Country A—yes, very active, all new active ingredients encouraged to be submitted as joint reviews.
b. Country B—not discussed but likely less of a focus than informal regional MRL interactions and MRLs focused on their crops for export. Resources could be an issue.
c. Country C—yes in toxicological joint reviews but not in MRL joint reviews for two reasons—the discrepancy in number of trials which hinders their ability to contribute and resource (labor and language) issues.
d. Country D—yes, have participated in a number of global joint reviews under the OECD framework that result in usage approvals and establishment of associated MRLs. Active supporter.

5. Import/export issues/priorities varied by each respondent, depending on their agri-food trade status, dependencies, and strategies. Many deferred answering, referring questions to other government departments.

a. Country A—this is not the respondent's area of regulatory expertise.
b. Country B—highly export-focused and want to ensure that they do not have MRL violations when shipping to their destination markets. MRL focus is on their export markets, MRLs on imports less of a priority. Shipments destined for Country B should defer to Codex MRLs as best option. Very concerned about MRL transitions to positive list in South Korea as well as problem of older generic products with multiple registrants which either lack data or resources to submit a MRL application to South Korea. They propose many of their national MRLs at the levels of their important export markets.
c. Country C—have faced MRL rejections abroad due to either the MRL being more stringent in the destination market or not approved. They face the same measurement discrepancy in field trial data again, should they try to seek an import tolerance in

that destination market. Their producers can amend use of pesticide products when an issue arises in a destination market.

d. Country D—deferred to the appropriate government departments and agencies for details. Did mention that importers and food manufacturers that rely on imported ingredients can address potential adverse impacts from imported food that is non-compliant with their MRLs by applying to have the relevant agency consider harmonizing their MRLs with those permitted in the imported product's originating country.

6. Suggestions on MRL facilitation/harmonization

   a. Country A:

      i. The main issue with Codex is timing as it requires the new active/new use to be registered within at least one country (that has their own regulatory system) prior to being submitted to Codex.
      ii. Encouraging joint reviews with major trading partners is the most efficient pathway. If the countries involved are using the same data packages during their initial review (whether new active ingredient or new crop MRL, it is more efficient to establish a harmonized MRL from the beginning rather than try to resolve differing MRLs after they are set).
      iii. Countries with their own regulatory review systems (generally OECD members) have established processes for joint reviews, one outcome of which is the higher likelihood of harmonized MRLs from the outset. It is much more difficult to establish these processes with countries that do not have a recognized review system, or, either do not operate under Good Laboratory Practice (GLP) or are not recognized as GLP compliant. As all pesticide residue data in OECD countries must be GLP compliant, this is a big hurdle to the acceptance of other countries data.

   b. Country B—concern about older generic pesticides not having sufficient data or resources to submit MRL applications in various markets. Focus is on establishing MRLs for priority crops destined for export and works with the pesticide registrant when approving domestic registration for use on their crops.

c. Country C:

i. They liked the idea of harmonizing data package requirements, so you could use one data package to get MRLs in many markets. They mentioned this was discussed at the OECD. It is also the basis of an APEC paper[3].
ii. Standardizing the number of field trials was again mentioned.
iii. Having a database with minor uses along with a data package might be helpful for reviews.
iv. They also said that there was a real issue with developing countries needing MRLs for their products like tea and tropical products. They do not have the resources to pay for the MRLs and may not have the data. Perhaps, something could be set up to assist with the issue, so there are fewer violations for them.

d. Country D:

i. Strongly supports the efforts of the APEC FSCF in facilitating voluntary regulatory cooperation work in the area of import pesticide MRLs in the APEC region, particularly the facilitation of trade in agricultural commodities through the development of a procedure that economies may apply for in the establishment of import MRLs.
ii. Supports the APEC Guide as a voluntary aid for APEC economies when considering the alignment of domestic pesticide MRLs of plant protection products, with relevant international or regional standards (e.g.., Codex, ASEAN MRLs) or an exporting economy's MRLs.
iii. Notes that WTO members have the right to adopt SPS measures to achieve their appropriate level of protection. Their country develops MRLs based on scientific risk assessments and is applied in a consistent manner and recognizes that MRLs may be set at different levels due to a range of factors, in particular, different GAP.
iv. They support in principle, that as much as possible, MRLs should align with relevant international standards such as those established by the Codex. They set their MRLs, associated with registered products and approvals, based on

rigorous risk assessments. The MRLs are for agricultural and veterinary chemicals approved for domestic use and others considered for import tolerances. Their process for developing MRLs is transparent and open to external comments.

Discussions were also held with two US-based policy makers—one involved in the IR-4 project[4], the other in the EPA. Their insights include:

1. The OECD calculator is a highly useful tool in helping establish aligned MRLs. Some issues that need to be addressed include:

    i. Differences in the data entered will result in different MRLs—how the measurement is made, the methodology, definitions, exclusions, etc. Even between close trading partners such as US and Canada, the way data are defined or obtained differs, resulting in different values that are then entered into the calculator, and a divergent MRL can arise.
    ii. The calculator is also not yet capable of accommodating crop groupings—these are being worked on.
    iii. Asian countries are generally unfamiliar with the calculator, limiting its universal acceptance. Language, technical, and labor capacity—in terms of ability and availability and overall resource constraints—are probable causes.

2. That Codex becomes more accommodating of countries' unique conditions in joint data package submissions, and by doing so, focus less upon local level practices as criteria to establish a MRL standard. This would provide greater flexibility to accept joint submissions and data packages from country groups and establish MRLs that would enable trade and, hence, progress towards harmonization. For example, three countries jointly submit 20 data packages to the JMPR for approval of a MRL for a compound used on rice. Due to their unique differences in pests, application rates, growing seasons, soil, climate etc., they each have different GAP for that compound for its use on rice; they may also have revised their respective GAP to reflect new data, but the domestic label for the product remains the same. JMPR may note that of the 20 trials submitted, only six match the GAP exactly and reject the submission, and no MRL will be established. They will ignore

the remaining 14 trials despite their scientific validity. Yet, the data are individualized for each country's respective needs and these differences are what cause the discrepancy. Instead, JMPR should consider all the data provided to them, not just those that fit into narrowly defined parameters such as the domestic GAP or label. Currently, what is approved domestically, due to country-specific conditions, precludes the combination of data packages from multiple countries in a joint submission, to achieve a harmonized MRL; the asynchronous data result in rejection of the MRL application. JMPR should focus on determining that the trials and data are valid and that the proposed MRL would pass their risk assessment. JMPR would essentially establish a MRL at a trade standard and the local level data and enforcement would be left to each country's respective government. Providing greater flexibility in accommodating country-specific conditions would allow more countries to submit more data and facilitate the establishment of more MRLs.

3. Joint or even regional reviews—countries with trade relationships or like-minded goals should make simultaneous and concurrent MRL submission applications—apply at the same time with the same data package—is a proactive step towards harmonization and easier than trying to harmonize after the registration or MRL is approved.
4. Problems with Codex are capacity and timing.
5. Developing countries are increasingly aware of issues—better to choose Codex and work collaboratively on communication, cooperation and awareness.
6. Asian countries or region—setting their own MRLs, because civil society groups are driving politicians to protect safety of food supply.
7. Engagement is key.
8. Maintaining focus on risk assessment at the various forums working on MRLs is essential; otherwise, the agenda becomes cluttered with tangential or minor topics.
9. Increasing the functionality of Codex is critical—via increased resources, including experts, knowledge, and finances, as well as reforming procedures and processes to increase efficiency. Set schedules and timelines, such that procedures do not block movement forward.

The interviews show that interest in the MRL issue is high, but distributed according to the priorities of individual countries. Exporting countries focus on ensuring that their exports do not experience violations in important destination markets. Developed countries with the resources to do so participate in international MRL efforts such as joint reviews but even so, still face barriers in language and resource scarcity such as having sufficient technical personnel. Technical differences in data parameters such as in number of required field trials, analytical methods, definition of residue, parent compound or metabolite, for example, also pose a challenge for countries, even developed ones. Of interest is that developing countries will work with the pesticide registrant to develop domestic MRLs, a potential indicator that lack of resources and technical capacity remains a significant issue. Developing countries are aware of and monitor developments in global MRL issues but focus their limited resources on developing MRLs of priority to their domestic production and exports; for example, they lack capacity and resources to participate in joint reviews. Crop groupings are of interest to respondents, but again, participation is limited by lack of resources and capacity. These findings reflect and are consistent with many of findings in the academic literature.

Notes

1. The OECD calculator is a highly useful tool in helping establish aligned MRLs. It allows for standardized data input among countries when establishing MRLs. It is described more fully in the next Section.
2. The Joint Meetings on Pesticide Residues is one of the Codex bodies responsible for developing MRLs. The JMPR consists of independent FAO and WHO scientific experts who review and analyze data on pesticides.
3. The APEC Food Safety Cooperation Forum (FSCF) guideline which is discussed later in this book.
4. The IR-4 Project is a collaborative effort between the US EPA, USDA and the ICGCC to facilitate the registration of minor use pesticides through the collection of residue and efficacy data.

CHAPTER 8

Institutional Initiatives to Deal with Internationally Asynchronous MRLs

Abstract The current initiatives by international institutions to deal with asynchronous regulatory regimes for MRLs are explained. The organizations examined include the Codex, the WTO, APEC, and the OECD among others. Despite considerable efforts in these international fora, progress remains a slow and resource intensive process.

Keywords Codex Committee on Pesticide Residues · Joint Meeting on Pesticides and Residues · Data submissions · Crop groupings · IR-4 · SPS Agreement · Comment system · Level of detection · APEC Food Safety Cooperation Forum · OECD calculator · OECD global joint reviews · Fulfillment costs · Misaligned incentives

Industry, governments, and international institutions recognize the increasing problem that heterogeneous MRL regulatory systems pose to the international movement of agri-food commodities, risking global food security, and increasing costs to both consumers and producers. That the majority of MRL violations occur because of misaligned MRLs is also commonly accepted. Acknowledging there are many factors contributing both to the persistence of MRL misalignment, as well as impeding efforts to ameliorate them, there have nonetheless been substantial and ongoing efforts to address this complex and technical issue within domestic governments, bilaterally, regionally, and multilaterally.

M.T. Yeung et al., *Declining International Cooperation on Pesticide Regulation*, Palgrave Studies in Agricultural Economics and Food Policy, DOI 10.1007/978-3-319-60552-4_8

One of the challenges to these multiple efforts across various fora is bringing them together into a focused, cohesive discussion that may facilitate a solution agreeable to all stakeholders. This section outlines the majority of efforts currently being undertaken in addressing MRL heterogeneity in order to understand areas of progress, potential for advancement, and opportunities for harmonization.

8.1 Adjusting Domestic Requirements

Domestic regulatory requirements have acted to prohibit MRL harmonization. For example, Japan previously would not allow a data application for an import tolerance to be submitted until the compound was fully registered elsewhere, with a label. Once that was completed, a data submission could commence in Japan. A Japanese review typically takes 18 months to 2 years to complete. The implication is that the earliest a new MRL could be established in Japan would be at least a year and a half after it had been registered for use elsewhere. This was frustrating to potential exporters who wanted to use new products and to registrants who sought to sell a new compound. In 2013, Japan adjusted its policy to allow for submissions to occur simultaneously with other markets. A MRL can now be applied for in Japan at the same time it is sought in the US, Australia, or the EU. Previously, the Japanese MRL would not be established until the product was registered elsewhere, but this new policy reduced by a year, the time between when a product is registered in a market and the establishment of a Japanese import tolerance. Canada has also streamlined its registration and review process, and the EU will consider Codex MRLs so long as there are no objections pertaining to the data raised. Australia has an annual system where foreign entities can request import tolerances using Codex or other MRL data, resulting in new MRLs within a year. All of these domestic steps have helped to address MRL differences.

8.2 Registrants Now Have Global MRL Strategies

Gone are the days where a pesticide registrant could introduce a new crop protection product to market without an international MRL plan. Potential registrants are highly cognizant that when a new compound is ready for release, the first question to arise will be what the MRLs in X, Y, and Z market are. As a result, they have created international

MRL strategies to obtain tolerances in key export markets and at Codex. Although these MRLs may not be established simultaneously, they can be in place in a relatively short period depending on the market's approval systems.

8.3 Codex Alimentarius Commission

The Codex Alimentarius Commission (CAC), the United Nation's (UN) international food safety organization supported by the World Health Organization (WHO) and the Food and Agriculture Organization (FAO), maintains a list of pesticide MRLs. These Codex MRLs are intended to serve as international MRL standards. They are used by some countries as their official MRL list. Others use Codex MRLs to supplement limited national lists. More recently, with the trend toward national positive lists, Codex MRLs are frequently adopted into the national lists, such as in the EU and more recently by China. They also frequently serve as the basis for MRLs during regulatory transitions to a positive list. The data packages for Codex MRLs are also referred to when countries establish their national MRLs. Hong Kong and Japan both started their expanded lists by initially using Codex levels. Codex MRLs play an important role in international MRL harmonization and addressing MRL challenges.

The two relevant bodies for setting MRLs are the Codex Committee on Pesticide Residues (CCPR), and the Joint Meeting on Pesticides Residues (JMPR). The JMPR consists of independent FAO and WHO scientific experts who review and analyze data on pesticides against a range of criteria to determine what the safe residue level should be on a crop grown under GAP, and what the safe daily intake of any residue would be for human consumption. Once the review has been completed the JMPR and then recommends MRLs to the CCPR.

The CCPR consists of representatives from national governments, and, while the JMPR is an independent scientific process, country-specific interests can come into play through the CCPR process. The CCPR follows the CAC 8-Step Procedure as the pathway to having a MRL adopted and is designed to ensure that a thorough scientific review and analysis has been completed in order to establish a MRL. It also provides member countries ample time to study the JMPR recommendation and provide comments. This has been criticized as being an overly long and slow process which can delay the adoption of the MRL or, in fact, preclude adopting the MRL at all. This process had been known to require a decade.

In the last decade, several important Codex reforms have occurred to improve efficiency in the Codex MRL setting system. Beginning in 2006, the CCPR revised the eight-step procedure to reduce the time needed for a standard to be evaluated and adopted, resulting in a significant increase in the number of pesticide MRLs being adopted. MRL adoptions can now occur in 2 years or less. The increased use of electronic working groups has also helped to advance inter-sessional work between the annual meetings. Objections by member countries could be noted, but unless there was a serious question about the safety of the recommended MRL, the MRL would advance to final approval.

Codex has also standardized how and when reviews are conducted. A formal process has been adopted with concrete deadlines on how a MRL can be requested through the Codex system. This has streamlined the process, such that registrants and other stakeholders are aware of deadlines for data submission and what year a compound is subject to review or reevaluation.

Another important step in the Codex process, as well as in other harmonization efforts, has been the move toward approval of MRLs by crop groupings rather than by individual crops. Crop groupings increase the scope of commodities covered under Codex MRLs. Crops sharing agronomic similarities can be grouped together and the data used in establishing a MRL for one crop can be considered for the entire group. This allows for more efficient use of scarce resources while extending MRL coverage to many more commodities. The data submissions used for major crops can be used for minor crops, as well; minor crops often lack the economic incentive (i.e., future profit potential) to incur the costs of preparing a submission to obtain a MRL.

A major challenge with crop groupings, however, is the wide variances in a single crop or grouping exhibited from one country to another for several different reasons including differences in climate, geography, diet, and economic environment, among others. Therefore, when work on establishing an international MRL through Codex is underway, achieving agreement on a specific crop grouping may slow the process. Despite this complication, essential efforts to harmonize crop groupings globally to the greatest extent possible are being made; the IR-4 Project has been the focus of efforts between the US Environmental Protection Agency (EPA), the Canadian Pest Management Regulatory Agency (PMRA), and the International Crop Grouping Consulting Committee (ICGCC) towards that goal for several years. Through revising and harmonizing crops within a group, crop groups can be adopted by Codex. All countries are then encouraged to adopt the Codex crop groupings rather than establish their own domestic groupings. This accelerates the establishment of MRLs for more crops, increasing the scope of MRL coverage across more products while increasing the efficiency of the overall process.

The capacity of the JMPR, however, continues to be an issue that members have tried to address. With the growing number of MRL submissions, increasing demands are made against the limited availability of human and financial resources, hence setting priorities for reviews can be difficult. Some important steps have been made at Codex to expedite the establishment of MRLs; however, more work needs to be done to modernize the process. This would include procedures to efficiently choose priority MRLs to review, the means to effectively include more crop groupings and methods of ensuring consistency in data measurement and submissions. Member countries could also consider additional funding and adding an additional regular meeting of the JMPR.

The trend of countries moving away from adopting Codex MRLs and implementing their own national MRL lists may well affect whether or not national governments would be willing to direct more resources towards Codex standard setting. Furthermore, the WHO and FAO, which have faced ever growing restraints on their budgets, are prone to give priority to the treatment of human disease as well as animal and plant health-related incidents of economic significance such as avian influenza; MRLs are, at this point, simply less of a priority.

The availability of human and financial resources will likely continue to be a struggle for the FAO and WHO, but policy changes within the Codex process could compensate for the lack of new resources directed at devising MRLs. These include greater use of data from national reviews; aggressively harmonizing crop groupings for reviews; changes to the label policy that requires particular information before a MRL can be established; and maximizing the use of all available communication technologies for inter-sessional work.

The capacity of the JMPR to undertake MRL assessments will need continuous attention; however, delays can also occur in the CCPR as the adoption of a MRL proceeds through its 8-step procedure. The CCPR sets the MRL priorities for JMPR consideration; this is a lengthy procedure and is often subject to the national interests of certain countries.

MRL establishment for minor use pesticides or those for use on specialty crops can face long delays as they may be of a low priority for some members. Further examination and refinement of the criteria for setting priorities with a view to expediting the list could be considered by the CCPR.

Delays are commonly caused by differences between the JMPR scientific assessment and national risk assessments carried out in member countries. When this occurs, countries tend to give more credence to their national assessments and may completely disagree with the JMPR's work. The usual means to reconcile this discrepancy is that the CCPR and its country representatives will refer the MRL recommendation back to the JMPR and ask that additional scientific elements be looked at, request a reevaluation of the original conclusions, answer questions, and provide explanations. Depending on the amount of work, the JMPR, with its capacity limitations, may need considerable time to respond. Similar to priority setting, further examination and refinement of criteria to ameliorate scientific discord could be considered by the CCPR.

Issues of contention can delay progress on adopting new standards and can even preclude the adoption of a MRL. Examples of such issues include different views on the science and the interpretation of the scientific findings or the regulatory procedures used to arrive at a conclusion. Others are a result of differences in national food policies which are usually driven by domestic civil society pressure or politics. The CCPR tends to avoid those MRLs with contentious issues and prioritize those with the greatest chance of agreement. It is incumbent upon the CCPR to promote mechanisms, such as task forces or working groups that would facilitate consensus building.

Despite several improvements, challenges remain at Codex, especially regarding its capacity to conduct reviews. The Codex review agenda has reached its capacity limit, leading to delays of at least 3 years due to a limited number of reviewers and scarce financial resources.

8.4 Inter-Regional Research Project Number 4 (IR-4)

The IR-4 Project is a collaborative effort between the US EPA, USDA, and the ICGCC to facilitate the registration of minor use pesticides through the collection of residue and efficacy data. Minor use crops are those that have a high production value but are not widely grown. As a result of their smaller production volumes, the registration of crop protection products for use on these crops is usually bypassed as the costs of registration far outweigh the registrant's return on their investment.

The data collected through the IR-4 provide the necessary information for regulators to make a decision allowing the use of a chemical on minor use or specialty crops. The initial role of the IR-4 Project was primarily to assist regulators in the US by providing data on minor use pesticides; its role has since evolved to become an important international partner in MRL harmonization. It collaborates with, provides data and reports to, institutions working on MRL harmonization such as Codex and the OECD. It also partners with other governments to advance work on minor use pesticides. The IR-4 has worked closely with the PMRA and Pest Management Centre of Agriculture and Agri-food Canada on the development of the Canadian Minor Use Program with the goal of creating a model that could be adopted by other countries.

8.5 World Trade Organization

If member countries concerned about trade disruptions due to the asynchronicity of MRLs are willing, the WTO, and its subsidiary SPS Committee, could provide a good forum for multilateral discussion and development of a strategy to move towards more global harmonization of MRLs. The text of the SPS Agreement contains numerous provisions that, if applied consistently by member countries, could significantly reduce trade disruptions due to the proliferation of national MRLs as well as those MRLs absent in national regulations.

The first relevant provision pertains to the adoption of international standards by countries into their domestic legislation. Codex is specifically referenced in the SPS Agreement as the international body recognized to establish food safety standards. A MRL established under the Codex, as discussed in the previous section, becomes the de facto accepted international standard.

The SPS Agreement does allow countries to set their own standards in order to meet the level of protection desired by the sovereign country itself, but in setting those standards, WTO members are required to base their standards on science, as well as being determined through a risk assessment. Furthermore, those standards cannot be more stringent than necessary to meet the country's level of protection. These provisions are in place to prevent countries from employing standards to serve as a means of economic protection against foreign imports for the domestic agri-food industry.

While the SPS Agreement does not obligate, but only encourages, countries to adopt international standards, it does say that those who do so are deemed to be in compliance with the SPS Agreement; no further scientific justification is required. It also ensures that a member which chooses to adopt and apply the Codex standard to imports should not be legally challenged under the WTO dispute settlement process. If a WTO member country chooses to use its own national standards rather than international standards, it can be challenged under the dispute settlement process should another member country believe that the national standard is not in compliance with the SPS Agreement (e.g., not being based on science).

The transparency provisions of the SPS Agreement require that countries give advance notice to the WTO of proposed changes to domestic standards, and to promptly publish those changes once they have been made. It further requires countries, upon request from other countries, to provide an explanation of the reason that the standard has been put in place or changed. For the introduction of a new MRL, these notification requirements give government officials, whose country's exports could be affected by the new standard, the opportunity to raise its concerns with the imposing government's officials who represent it at the SPS Committee meetings. The Committee meets three times a year, and members can formally raise their concerns in the plenary and ask for an explanation from the country in question, as well as have the potential trade impact outlined.

While this comment system is conceptually sound, some of its aspects have been criticized as being ineffective in practice. Countries may make their notification to the WTO quite late in the process of changing or establishing standards, leaving a truncated comment period for affected countries to respond. By the time that comments can be made, the regulatory change is already completed. WTO notifications can also occur, when de facto, countries have already implemented the change. In addition, if comments from affected countries are permitted, they can also simply be ignored with no response forthcoming. The change in a standard can be semantically classified as something other than *new* or anything other than what would require a notification to the WTO; hence, no WTO notification is made, and affected countries do not have notification of the change or time to comment. As illustrated in the case studies, in 2013, the EU reduced the scope of the MRL for phosphonates from both a fertilizer and pesticide to only that of a pesticide as a form of enforcement action rather than a change in EU legislation. Consequently, no formal WTO notification was issued to inform third country stakeholders of the change. Hence, no opportunity was given to comment on unintended impacts, leaving only a limited timeframe to achieve compliance or assemble the required data submissions from third country stakeholders (Vandercammen 2016). The comment system requires some adaptations to make it more practical and effective in its implementation.

In the case of missing MRLs that are impeding trade, the SPS meetings also provide a chance for member governments to formally and publically encourage other countries to abide by the intent of the SPS Agreement and adopt the Codex MRL, either permanently or at least during the transition period, while its own MRL is being established. The physical meetings also provide an opportunity to build alliances with other like-minded countries sharing similar MRL concerns. The meetings also provide a forum, whereby informal interaction between delegates can also contribute towards solutions. Should the SPS Committee not want to take on the MRL issue or provides only a very limited response to the matter, there is nothing precluding an informal group of willing countries from collaborating to advance solutions on the margins of the meetings. In an ideal world, the transparency provisions and the in-person meetings should give ample opportunity to resolve the matter; however, some countries may not notify the new MRL to the WTO, or refuse to work toward a solution due to domestic pressures.

MRL discussions arise from time to time in the SPS Committee, but usually occur reactively in the context of a country and commodity specific issue. At the March 2015 meeting, India tabled a paper to initiate deliberations on what it sees as a growing problem in international trade—the application of Levels of Detection (LoD) in importing countries. The paper describes some issues that India has encountered with its own exports because of LoD. The paper also reviews the provisions of the SPS Agreement that members are required to abide by when setting their LoD. Finally, it recommends that the systemic issues arising from the use of LoD should have further consideration within the SPS Committee. Finally, the paper outlines a few proposed solutions for discussion (Government of India 2015).

The Indian paper was not fully supported by all WTO members; it did, however, open a door for further discussion at the WTO. At the October 2015 SPS meeting, the Chairperson suggested the subject be further explored in a thematic session. The SPS Secretariat circulated a draft program of the thematic session for discussion at the March 2016 meeting. A MRL workshop was held at the October 2016 meeting, showing that the issue is increasingly being recognized as being a major trade issue. The SPS meetings will likely remain an important forum for discussion of MRLs and present valuable opportunities for international collaboration and discussion. Government representatives could be encouraged by industry to not only fully engage in and prioritize the process, but to use the SPS Committee process to develop a way forward.

8.6 Asia-Pacific Economic Cooperation (APEC)

The APEC Forum attempts to strengthen economic co-operation amongst 21 regional economies by cooperatively removing impediments to trade, both at the border and inside the border. The goal of APEC initiatives is to reduce the cost of doing business by providing trade information, and simplifying processes faced by firms active in the region. APEC also strives to help economies build capacity and, in doing so, it partners with organizations such as the World Bank and with the private sector.

In 2007, APEC created the Food Safety Cooperation Forum (FSCF) with the goal of encouraging economies to work together in harmonizing food safety standards with international standards, as well as to create greater transparency among member economies. In 2013, the FSCF agreed to an action plan to further facilitate regulatory cooperation in the region. An important part of the plan was to develop a roadmap for MRLs to include adoption of Codex standards; recognition of other economies' MRLs when one does not exist in the country of import, and share work and exchange data to establish MRLs. A working group was established to develop the roadmap.

APEC has undertaken two efforts pertaining to MRLs. The first, launched by wine industries around the Pacific Rim, seeks harmonization of wine grape MRLs in markets around the world. This effort is headed by the Wine Regulatory Forum (WRF), a group of regulators and industry stakeholders from the 21 APEC countries. Established in 2008, the

WRF is tasked with coordinating and harmonizing international wine regulations in the region, and where possible, the exchanging of information and increasing capacity-building efforts among members. A key part of this effort focuses on harmonizing MRL regulations. Since 2008, APEC wine industry members have tracked the pesticides used in the production of wine grapes, and tested product to determine the levels of pesticide residues in wine, all in an effort to determine which compounds are MRL priorities. In 2014, WRF members agreed to identify a priority list of wine grape pesticides for which Codex MRLs should be sought. The WRF is representative of an industry-led effort to coordinate pesticide regulations and jointly develop MRL priorities among multiple countries.

Subsequently, a second effort, the *Import MRL Guideline for Pesticides* (APEC 2016), was released for comment in December 2015 and finalized in August 2016. It provides guidance to APEC member countries on how to standardize their evaluation methods to ensure MRLs are set in a timely manner, using similar processes. The guideline can be used by member economies, on a voluntary basis, to avoid trade disruptions caused by disparities in setting national MRLs. The guideline aims to,'where practical and appropriate, also achieve greater regulatory convergence of MRLs, promoting greater alignment with international standards, while reducing the regulatory burden across APEC economies and facilitating trade' (APEC 2016). To increase consumer confidence, promote greater regulatory alignment with international standards, and enable the sharing of MRL trial data and information amongst members are also objectives of the guideline.

A key component of the guideline is to facilitate greater clarity and alignment specifically on the application of assessment methodologies used when considering import MRL requests, particularly from the perspective of consumer protection. The guideline is voluntary and meant to be used as guidance for APEC economies seeking to align domestic MRLs with relevant international or regional standards such as Codex or ASEAN MRLs, or with an exporting economy's MRL. It is intended to provide a consistent framework for APEC economies to identify the minimum required data set upon which to assess an import MRL based upon health-based guidance, authorized use, and established MRLs, either internationally or in the exporting country.

The guidance document provides a step-by-step process for developing a MRL. It provides information on the MRL process, information required to support a request, the provision of data and examples of the application process. It seeks to provide practical guidance for countries on steps needed to establish MRLs, even though the regulatory review process may differ from country to country. The goal is to allow a single data package to be submitted to multiple markets. Discussions regarding implementation of the guideline continue, with a meeting scheduled in February 2017. Such an effort represents a positive step toward addressing the MRL asynchronicity.

While APEC is a regional organization, the development of the guideline presents an excellent opportunity to advance MRL harmonization not only within the region, but also globally. There is an opportunity to work with the FSCF and other fora such as the WTO SPS Committee to create awareness and promote use of the guideline.

8.7 Organization for Economic Development and Cooperation (OECD) Calculator

A practical tool that has contributed to MRL harmonization is the OECD MRL calculator. This calculator seeks to harmonize the data packages used to establish MRLs among countries. As opposed to reviewers in different countries looking at the same data yet arriving at differing MRLs based on a variety of factors, the OECD calculator allows reviewers to insert the relevant data and produces a MRL that can be used by all countries. This avoids unnecessary differences in MRLs. While all 34 OECD countries should use the calculator, it is unclear how many do in practice, and to what extent. It is also unclear if non-OECD countries are aware of and/or use the calculator. Some of the countries responding in the interviews provided in Chap. 5 indicate indirect use through import tolerances or no use at all.

The OECD MRL calculator harmonizes the calculation of MRLs across OECD partners by providing consistent data parameters to be used in the calculation of MRLs, based on pesticide residue measurements obtained from supervised field trials. It is largely based on other statistical methodologies for calculating MRLs used by countries, particularly those within the EU and the NAFTA region. In calculating a MRL for a particular pesticide use on a crop, a data set is established by measuring pesticide residue samples taken from field trials carried out according to the use stated on the pesticide label. The OECD Calculator sets a target by which most residues produced when following the label will be below the MRL. The OECD calculator has become accepted as providing statistically robust and scientifically defensible MRLs. The US and Canada, as well as some other OECD countries, use the calculator to better coordinate their pesticide regulatory frameworks and minimize trade disruptions due to misaligned MRLs.

8.8 OECD Global Joint Reviews (GJR)

In an effort to expedite and synchronize the registration of pesticides, including the establishment of MRLs at similar times across markets and to avoid having countries continually re-reviewing data that had already been evaluated by others, joint reviews of new chemicals became common in the mid-2000's.

A GJR involves two or more countries working together to review the relevant data and scientific information with the objective of approving use of a chemical and concurrently, establishing a MRL. The establishment of a MRL by the GJR not only significantly reduces the likelihood of non-harmonized MRLs causing trade disruptions, but is also a means by which national regulatory bodies can divide the workload in setting the MRL, share resources as well as review and question one another's work. The GJR process allows each participating country to base their evaluation and review on the same data package. Each country, however, still makes a sovereign decision on what the national MRL will be.

While GJRs have increasingly become the norm for harmonization of MRLs, they have, so far, been used by a very limited number of national regulatory bodies. They have been used mostly by countries that have a well-developed regulatory system that is used to establish their own MRLs and which do not automatically accept Codex MRLs. They may consider the work of Codex, but ultimately, they establish their national MRLs through their own domestic legislation and regulatory process.

Countries rely heavily on their own domestic regulatory systems to establish MRLs, but also may value working with other countries in sharing field trial data and the related workload, usually when a significant amount of trade in commodities is involved between the countries.

Several countries have participated in these joint reviews, including the US, Canada, Japan, South Korea, Australia, China, Brazil, and EU member states. In recent years, the number of joint reviews conducted has declined, with fewer EU member countries participating. It has been suggested that differences in technical and regulatory capacity can act as a disincentive for some countries to participate (Pratt 2016). There were, however, more GJR occurring in 2016 than in recent years. Overall, dozens of chemicals have been reviewed through this program, resulting in harmonized MRLs being established in multiple markets over a shorter period of time. The OECD, however, is only a club of countries with a membership limited to, for the most part, relatively rich countries and does not have a full multilateral reach.

8.9 Fulfilment Costs and Misaligned Incentives

The trade problems and associated costs of forgone economic opportunities pertaining to the absence of regulatory harmonization is not a new topic nor is it confined to maximum residue limits for pesticides. The problem has long been recognized by governments, as well as industry and other groups in civil society. That governments have long recognized the problem is evidenced in the numerous multilateral organizations established to deal with the problems including the World Organization for Animal Health (1924), the International Plant Protection Convention (1951), the Codex (1961), the Agreement on the Application of Sanitary and Phyto-sanitary Measures of the WTO (1994), and the Agreement on Technical Barriers to Trade of the WTO (1994). There are also numerous pluri-lateral, regional, and bilateral forums and institutions. Thus, governments are cognizant of the problem at one level and have been willing to take action in attempts to address the problem. These organizations deal largely with reducing what have been called *fulfilment costs* (Kerr 1997) and are similar to private sector transactions costs (in that they represent friction which inhibits the smooth functioning of goal achievement (Hobbs 1996)—for example, international regulatory harmonization). The OECD calculator and the APEC Guidelines represent mechanisms to reduce *fulfilment costs* for countries having a goal of harmonization.

Despite these best intentions of politicians, regulatory harmonization has proved to be both frustratingly slow and elusive. One of the problems is that there are many technical standards, entailing a very large volume of work. Hence, governments must prioritize and getting priority status for ones' problematic technical standard requires considerable resources for lobbying and other informative activities—another aspect of fulfillment costs. Gathering sufficient resources for these activities often requires cooperation among groups with myriad respective interests in diverse standards, and it may also require a coordinated effort with like-minded groups in other countries. Given that removing regulatory divergence as a trade irritant often requires technical assessments to be undertaken over extended periods of time, momentum for change may have to be maintained over a considerable span of time. Maintaining momentum is difficult both domestically and internationally, given limited resources and the multiple responsibilities of individuals.

Those interested in removing regulatory divergence are normally motivated by the benefits that will come from the expansion of trade—increased profits are the incentive. Those charged with the development and implementation of regulations, however, often have no interest in the benefits of expanding trade, but rather have incentives tied to assuring human health, protection of the environment, or personal or political job security. In short, they have no positive incentive to countenance, much less to actively seek, the removal of regulatory divergence. Without a major institutional reform, re-alignment of incentives within national bureaucracies is not likely possible. This means that friction in the process remains and far more resources will be required for fulfilment to be accomplished. The misalignment of incentives within governments and the fulfilment costs associated with them have been explored at greater length in Kerr (1997) and Smyth et al. (2015). What is new to the discussion is the spillover effect on the ability to overcome future food security challenges.

One alternative is to create an independent institution that is not formed from individual ministries. The members of this institution can be technical experts. They should be charged with pro-actively developing mechanisms to remove regulatory divergence such as asynchronous regulations or the conditions under which equivalence can be granted. Their incentive should be tied to their success in removing regulatory divergence. Once their proposals regarding a specific regulatory change are developed, they should be presented to the range of ministries that have involvement. These ministries must be given an opportunity to respond but must also respond pro-actively if they object to the proposed change (or part of the proposed change). If they do not pro-actively respond, the changes will proceed and begin the process to become regulation or legislation. This suggestion is loosely based on the model of the European Commission—the EU has had considerable success in regulatory harmonization—where the Commission is composed of individuals that are independent of individual governments.

The Commission has been given the sole responsibility for developing proposals to remove regulatory barriers, and once they have been developed, they are presented to the Council of Ministers and the European Parliament, which must pro-actively object if a proposal of the Commission is not to be adopted. If a proposal is rejected, it reverts to the Commission to develop a new proposal. Hence, the development of proposals is removed from those lacking incentives to develop one.

Individuals with technical ability can operate independently of national governments. Lending expertise to organizations charged with harmonization can assist in reducing fulfillment costs (e.g., helping to develop the OECD calculator). It is possible to send those with expertise to work in international organizations but they are not to be the representatives of their individual countries. The staff of the WTO, for example, has often previously worked in government ministries but now work for the WTO, not their individual governments.

There are inklings from experience that such an approach can work. This pro-active activity was undertaken in a number of venues including the SPS committee of the WTO and the APEC FSCF. The latter produced the Import MRL Guidelines for Pesticides which has aspects that could form the basis for regulatory harmonization. There is no mechanism for formally having members assess the Guidelines and they remain, while widely considered useful, simply guidelines—but they do have an important role in reducing friction in the harmonization process (e.g., reducing fulfilment costs).

References

APEC. 2016. *Import MRL Guideline for Pesticides*. http://publications.apec.org/publication-detail.php?pub_id=1750. Accessed 17 Feb 2017.

Government of India. 2015. *Need for Measures on Detection of Pesticide Residues not Registered in the Country of Import for Unimpeded flow of Trade*. Communication from India, WTO SPS Meeting, 26 March 2015.

Hobbs, J.E. 1996. A Transaction Cost Approach to Supply Chain Management. *Supply Chain Management* 1 (2): 15–27.

Kerr, W.A. 1997. Removing Health, Sanitary and Technical Non-Tariff Barriers in NAFTA. *Journal of World Trade* 31 (5): 57–73.

Pratt, S. 2016. Joint Reviews of Pesticides Encounter Roadblocks. *The Western Producer*, July 14. http://www.producer.com/2016/07/joint-reviews-of-pesticides-encounter-roadblocks/. Accessed 15 Feb 2017.

Smyth, S.J., P.W.B. Phillips, and W.A. Kerr. 2015. Food Security and the Evaluation of Risk. *Global, Food Security* 4: 16–23.

Vandercammen, G. 2016. *New EU MRL for Fosestyl Takes Effect*. GAIN Report #E16001, Foreign Agricultural Service, Washington: United States Department of Agriculture, April 1. https://gain.fas.usda.gov/Recent%20GAIN%20Publications/New%20EU%20MRL%20for%20Fosetyl%20Takes%20Effect_Brussels%20USEU_EU-28_1-4-2016.pdf. Accessed 15 Feb 2017.

CHAPTER 9

The Way Forward

Abstract A number of suggestions arising from the insights gained in this research are presented. They represent practical means for achieving regulatory harmonization for MRLs. They will require both the political will to move forward and a spirit of international cooperation. Enhanced food security for the world's poorest consumers should provide a motivation for both.

Keywords Default · Joint reviews · Concurrent submission Parameters · Methodologies · Data · Research hubs Modernize · Funding · Staffing · Claw back · Pesticide food safety standard · Engagement · Advocacy · Relationship building · Outreach Collaboration · Proactive · Friction

There are two components to the problem of misaligned global MRLs. The first is how to garner the support and engagement of government regulators and officials to take on the complicated, technical issue of MRL harmonization. The second is what actual regulatory or policy changes that move towards MRL harmonization can be made. The issue of MRL misalignment has been described as *death by 1000 cuts*; the solution will be by 1000 small steps as a *silver bullet* answer is unlikely. The concern regarding upcoming food security challenges adds urgency to what has heretofore been perceived as solely a trade problem.

M.T. Yeung et al., *Declining International Cooperation on Pesticide Regulation*, Palgrave Studies in Agricultural Economics and Food Policy, DOI 10.1007/978-3-319-60552-4_9

The literature review, industry discussions, and interviews contributed towards several suggestions for reducing trade uncertainty due to MRL misalignment:

1. If a MRL is missing or absent in a destination market, default to the exporting country's MRL or Codex MRL rather than to LoD or zero tolerance. This is a key recommendation in the APEC Import MRL Guideline for Pesticides and will do much to reduce trade uncertainty if followed. Adopting or following as much as possible, any or all aspects of the APEC Guideline deserves serious consideration.
2. Countries with trade relationships or like-minded goals should conduct joint MRL reviews and make simultaneous and concurrent MRL submission applications. By using the same data packages during their initial review (whether new active ingredient or new crop MRL), it is more efficient to establish a harmonized MRL from the beginning rather than try to resolve differing MRLs after they are set. Conducting joint initial reviews will prevent MRL misalignment.
3. Harmonize the parameters and methodologies used to generate and gather data to be used in the MRL application submission or the OECD calculator. Do so while considering and allowing for regional and local differences in pesticide GAP and use. This prevents technical differences in data from causing misaligned MRLs.
4. Promote, create, and support global regional MRL research hubs. The hubs would operate on a consistent procedural and data framework. These can then share the workload, costs, resources, and data needed for MRL submission packages. Regional hubs can also then accommodate language issues, share staff, and capacity, as well as regional knowledge pertaining to crops of interest and of similar applications. For example, if a MRL is needed for mango, four regional hubs in North America, the EU, Asia, and South America would then conduct an agreed upon number of field trials. This would increase product MRL coverage and increase data sources. The countries participating in the

hub could then make simultaneous concurrent MRL submission applications. Within each regional hub, mutual recognition agreements or equivalence could be implemented to accommodate individual country priorities or needs.

5. Enable the OECD calculator to accept crop groupings and harmonize parameters for data generation/collection as suggested in point 3 above to ensure that the same data are being used. Build awareness regarding the calculator amongst Asian and non-OECD countries as an avenue towards addressing asynchronicity in MRLs.
6. Modernize Codex on a number of fronts. Increase funding and staffing to increase its capacity and speed the MRL process. Streamline procedures and processes, such that comments, questions, replies, or objections cannot act as indefinite blockages to the establishment of MRLs. Establishing set timelines and periods for steps in the process may assist this. Ensure that Codex has access to the best experts and information possible. Reform their review procedure to be more accommodating of country-specific data in multi-country submissions.
7. Countries are implementing divergent MRLs based upon often non-transparent reasoning. As yet, a dispute has not been brought to the WTO regarding MRLs acting as barriers to trade. Such a suit, if pursued, might clarify the basis upon which MRLs should be established and raise the profile of the problem, such that greater engagement by governments would ensue.
8. Increase the practical applications of the notification and comment system for MRLs at the WTO, such that:
   (a) Members cannot eschew notifications based upon semantics or technicalities; notifications should allow for sufficient transition periods for affected third countries to adapt.
   (b) Members should ensure that their WTO notifications process makes allowances for comments from affected countries and incorporates these comments early in the regulatory review to accommodate any potential adjustments.
   (c) Members must address objections or comments rather than ignore them.

(d) Establish set timelines for periods of comment, review, or objection, as well as for responses to such to ensure that while input is allowed, the process itself does not become an obstacle.

9. The APEC Guideline could be used to establish some form of accountability for ensuring that MRLs are established using risk assessment and attempt to harmonize as much as possible.
10. One suggestion to address the concern that generic or older pesticide products lack MRL applications would be to mimic the IR4 model at regional hubs to focus on MRLs for generic products or use JMPR Codex data.
11. As per Winter and Jara (2015), work towards the conceptual *claw back* of MRLs as less an indicator of pesticide food safety standards, and more as an indicator of GAP. Promote/increase the acceptance of the concept and approach of PFSS as the indicator for the food safety of pesticide residues. MRLs and PFSS could work jointly in concert.
12. Work to highlight and build awareness of the linkage between international harmonization of MRLs and food security to raise awareness of the issue among policy makers.

Achieving the engagement of governments to treat MRLs misalignment as a priority issue is less about policy and regulatory changes and more about consistent advocacy, relationship building, outreach, and collaboration.

When dealing with technical trade issues, countries are far more reactive than proactive. A reactive approach to trade disruption is less desirable as mitigation ex-post is highly resource intensive; governments and the affected industry of the exporting country spend a significant amount of time, human, and financial resources to re-open a market. When a shipment of perishable agri-food products is stopped at a port of entry, the costs can be exponential, to the individual exporter coping with an impounded shipment, and often, to the entire sector or industry which has lost a market for its products. Even when the technical issue is resolved, the long-term damage in the market can continue for many years, increasing costs for both exporting producers and importing consumers. While many trade disruptions are unexpected, in retrospect, there is often some form of proactive action that could have prevented or lessened the severity of the incident but was not taken.

Taking a proactive approach to mitigating trade issues such as MRL misalignment requires the consistent monitoring of trading partners' policies and actions as well as of global trends that could potentially impede future agri-food exports. Affected industries must identify and articulate the problem, including potential negative impacts and consequences, in order to enlist government support and action. This is not a simple task as governments tend to look at the expenditure of human effort and financial resources in comparison to the potential for a successful outcome. Other issues have ongoing priority and government officials can be reluctant to add another issue to their agenda. Other issues may also be answerable to greater political pressure than others.

While the above is not new, it is not *rocket science,* and only skims the surface of the work required. It forms, however, the essential backbone of how governments work together to advance international trade. Government to government, government to industry, and industry to industry connections are the foundation upon which to build.

In the context of global food security, however, this study points to one overwhelming conclusion. There is no substantive issue pertaining to the potential to harmonize MRLs. Given the care taken at the international level when developing harmonized MRLs, food safety does not appear to be at risk—and, of course, that should be the overwhelming priority. The remainder of the factors identified as inhibiting harmonization are simply friction, which can be removed or significantly reduced through sufficient political will and resources. While other asynchronous problems in agriculture, such as barriers to trade in genetically modified organisms, have received considerable attention (Isaac and Kerr 2007), MRLs have not. Given the almost universal use of pesticides and other chemical compounds in agricultural production globally, the effect of asynchronous MRLs on trade has the potential to cause major disruptions. The failure to address the issue is literally *frittering away food security.*

9.1 APPENDIX 1—Interview Guide and Questionnaire

Several agricultural groups are investigating potential strategies to reduce trade uncertainty related to the global asynchronicity of MRLs. With pesticide and chemical standards, regulations and MRLs varying from country to country, costly technical barriers to trade are created, impeding the movement of agricultural commodities as well as chemicals. This issue affects both importers and exporters of such goods, increasing costs, reducing competitiveness, and overall welfare for producers and consumers, in both developed and developing countries.

Individual countries undertake policy and regulations regarding pesticides and chemicals to protect their citizens' health, environment, and public safety. These are legitimate policy goals of any government. Hence, any effort to address the complex and contentious issue of MRLs must include the perspectives and priorities of individual country governments.

This short questionnaire is intended to support a larger study attempting to form a strategy in mitigating the impact that non-harmonized MRLs are having on global agricultural trade.

Thank you in advance for your time and assistance. A favor of a response is requested by March 18, 2016.

MRL QUESTIONNAIRE

1. Does your government use the international Codex MRLs?
   - (a) If yes, how?
     - i. If a MRL does not exist in your country, does your government automatically defer to Codex MRLs or use Codex MRLs to supplement a national list?
     - ii. Do you use Codex (JMPR) reviews or JMPR-recommended MRLs in establishing national MRLs?
   - (b) If no,
     - i. How does your government establish its MRLs, and why are Codex MRLs not being used?
     - ii. How long does it take to establish a MRL from time of petition?
     - iii. What would motivate/be an incentive for your government to defer to the Codex MRL if a MRL does not exist?
2. When establishing MRLs, does your government use the OECD MRL calculator?

   (a) Why or why not?
   (b) If not, what would motivate/be an incentive for your government to do so?
3. Does your government use crop groupings for MRLs?
   (a) Why or why not?
   (b) If yes, are Codex crop groupings being used?
   (c) If no, what would motivate/be an incentive for your government to use crop groupings?
4. Does your government participate in joint reviews for MRLs?
   (a) If yes, how active is your government?
   (b) If no, what would motivate/be an incentive for your government to become more active in joint reviews for MRLs?
      i. Are the joint reviews conducted through Codex or independent bilateralbilateral/multilateral cooperation (e.g., NAFTA minor use joint reviews)?
      ii. Do you have suggestions for improving the joint review process?
5. Have MRLs negatively affected your country's imports or exports? (please provide general details pertaining to product, market, and approximate value, if available)
   (a) How does/did that particular importer/exporter and/or industry cope with the MRL issue?
   (b) Did the MRL issue have broader impacts on other industries/sectors? How did these industries adapt?
6. Do you have suggestions regarding how MRL facilitation/harmonization can occur?
   (a) Suggestions for improving Codex's role?
   (b) Suggestions regarding WTO SPS?
   (c) Suggestions regarding regional forums such as APEC?
   (d) Suggestions regarding joint reviews?
   (e) Suggestions regarding other possible ways to encourage bilateral/multilateral cooperation in MRL-setting and harmonization efforts (through trade agreements, perhaps)?
   (f) Other suggestions?

References

Isaac, G.E., and W.A. Kerr. 2007. Whose Vision of the Future? The Entrenched International Conflict Over Genetic Modification. *The Geneva Post Quarterly* 2 (1): 87–107.

Winter, C., and E. Jara. 2015. Pesticide Food Safety Standards as Companions to Tolerances and Maximum Residue Limits. *Journal of Integrative Agriculture* 14 (11): 2358–2364.

Index

M.T. Yeung et al., *Declining International Cooperation on Pesticide Regulation*, Palgrave Studies in Agricultural Economics and Food Policy, DOI 10.1007/978-3-319-60552-4

The manufacturer's authorised representative in the EU is Springer Nature Customer Service Centre GmbH, Europaplatz 3, 69115 Heidelberg, Germany. If you have any concerns regarding our products, please contact ProductSafety@springernature.com

Printed and bound by CPI Group (UK) Ltd, Croydon, CR0 4YY
25/09/2026
02222890-0014